WITHDRAWN
NDSU

THEISM
AND
COSMOLOGY
GIFFORD LECTURES

THEISM AND COSMOLOGY

BEING THE FIRST SERIES OF A
COURSE OF GIFFORD LECTURES
ON THE GENERAL SUBJECT OF
Metaphysics and Theism
GIVEN IN THE UNIVERSITY OF
GLASGOW IN 1939

BY

JOHN LAIRD

Essay Index Reprint Series

BOOKS FOR LIBRARIES PRESS
FREEPORT, NEW YORK

224764

First published 1940

Published 1942 by The Philosophical Library

Reprinted 1969 by arrangement

STANDARD BOOK NUMBER:
8369-1147-4

LIBRARY OF CONGRESS CATALOG CARD NUMBER:
74-84317

PRINTED IN THE UNITED STATES OF AMERICA

Preface

THE present book is the first series of a general course on "Metaphysics and Theism". The second series, not yet delivered, is designed to have "Mind and Deity" for its theme. In the second series the provisional realism of the first series will be reconsidered. Pending this reconsideration, it seemed reasonable to postpone enquiry into such topics as divine tenderness and pity and righteousness. These subjects will be discussed (as I hope) in the later lectures of the second series.

To that extent the present book is provisional, but within its deliberately self-imposed restrictions, it is meant to be self-complete in the substance of it, and it is offered to the public in that spirit. Even if our present troubles had not made the future so much more uncertain than it normally is, I still would have preferred to publish the first series separately.

I should like to express my deep appreciation of the honour and the privilege of being asked to give these lectures by the Senatus of the University of Glasgow. I have now had two and a half years in which to become accustomed to the idea; but I still take a certain trembling pleasure in it. As regards my audience, my tremors have ceased and only the pleasure remains. I could not be shy in the presence of such friendly and of such attentive hearers, and I am accustomed to conceal my apprehensions from the reading public.

J. L.

KING'S COLLEGE, OLD ABERDEEN
November 1939

CONTENTS

LECTURE		PAGE
	Preface	5
	Synopsis of the Discussions	9
I.	*Concerning Natural Theology*	27
II.	*Concerning Theism*	55
III.	*The Cosmological Argument*	85
IV.	*Creation*	113
V.	*Eternity*	145
VI.	*Ubiquity*	174
VII.	*Omnipotence*	204
VIII.	*Teleology*	232
IX.	*The Argument from Design*	261
X.	*Examination of Cosmological Theism*	291
	Index of Proper Names	323

SYNOPSIS OF THE DISCUSSIONS

I. *Concerning Natural Theology*

THE older contrast between natural and non-natural theology was between philosophical and unphilosophical theology, the later contrast between a theology that was not and one that was based upon "revelation".

The ancients contrasted philosophical theology with mythological theology of the one part and with civil theology of the other part. In our view philosophy attempts to present its accounts ready for audit. Poetry and mythology do not. We therefore agree in principle with the ancients in this matter. We also agree that a ceremonial or ritual "theology" is just as impossible as a ceremonial or ritual arithmetic.

Regarding "revelation", it may be true that a Christian who knows the truth by "revelation" might be imprudent if he attempted to defend any of the "truth" on what might turn out to be the weaker ground of natural evidence; and parts of Christian theology are beyond the ken of "natural theology". But it would be sad if docility were the last word in this matter, if all theology was kerygmatic in the end, and left no scope for independent metaphysical enquiry.

It remains very disputable, however, whether the "natural" domain of theology, if there be any such domain, can be defined with any approach to precision.

The meanings of "nature" and of "natural" are very various. For our purposes the most important sense is what is called the *cursus ordinarius* of events, but there is here at least the suspicion of a circle. Are "ordinary" events those that can be ascertained by man's "ordinary powers", or are a man's powers called "ordinary" in proportion as they deal with "ordinary" events? In any case many theologians would hold that wilful self-restriction to what is "ordinary" or secular is a sort of metaphysical and theological treason. And many philosophers would hold that ultimate questions cannot be treated as if they were "ordinary".

Theism and Cosmology

A man's ordinary powers in this matter are sometimes described as his "reason" or, derogatively (?) as his "understanding". The description may be defended, however, if the use of vulgar logic is regarded as defining the level rather than as exhausting the scope of natural enquiry. A rough inventory of our natural powers would include perception, introspection, vulgar logical inference and perhaps some vulgar mind-reading (of the minds of our fellows). It would also include probability, analogy and a certain synoptic potency implied in all judicial balance.

Such an inventory can scarcely hope to be rigidly precise; but it is useful and we know pretty well what is meant by it. It is not plain indeed that we have any other effective knowledgeable powers than powers of this order, or that, if we have other powers, these are applicable to theology. The point remains obscure despite what certain theologians have said to the contrary, and despite certain philosophers who contemn, let us say, the "understanding".

The objection that these "ordinary" powers cannot be employed about questions that, like the origin of worlds, are *not* ordinary does not stand to reason, and may be more specious than solid. At any rate the problem is worth an enquiry.

If it be held that "religious experience" yields special evidence that is highly relevant to our theme, we have to reply that it has not been established that such evidence is not amenable to "the right use of ordinary means" and that it is legitimate to proceed upon the assumption that such special evidence does not yield a secure basis for philosophical argument.

In short, there is a place for natural theology even if a boundary commission about it would in some respects be perplexed to the end.

II. *Concerning Theism*

Anyone who asserts that there is a God or gods in any intelligible sense is a theist. This general sense of the term seems to beg fewest questions and so to be the most convenient. It follows that polytheism, pantheism and also deism should be regarded as theistic, not as anti-theistic theories.

We are concerned, however, with serious philosophical views upon this subject, not with any and every whimsical imagining

Synopsis of the Discussions

about godlings. Hence it seems advisable to be hospitable but not too hospitable towards serious theories that may reasonably be called theistic.

It is suggested that philosophical theism must be upon a cosmic scale and implies a certain unity in its conception of deity. With rather more diffidence it is further suggested that any developed form of theism must attempt to give an *ultimate* theory.

These topics are further debated. The God of a crude theism may not have been the God of the universe, but the God of developed theism must be at least diacosmic. Hence "humanity" or "culture" is not a possible candidate for divinity. It must also be understood that the power of deity is not weakly diacosmic.

Secondly, while certain varieties of the polytheistic pantheon may be childish, sound philosophy may be pluralistic as well as monistic, and a philosophical theology may seriously contend that deity is rather a spiritual or a super-spiritual community than a singularistic being. In other words, the unity that seems essential to philosophical theism (whether God be all or merely supreme) must not be interpreted in such a way as to prejudge the philosophy of the One and of the Many. This is highly relevant to the subject of pantheism.

The statement that developed philosophical theism always claims to be *ultimate* may be more perplexing. It is argued that the theory of a limited deity is not a denial of ultimacy; for the limitations, external or internal, might well be ultimate. Again, the theory of a "God in the making" may quite well be theistic, at any rate for those who believe that all existence is ultimately in process and, in that sense, "in the making." On the other hand the theory of the "emergence" of deity may be more doubtfully theistic. If, for instance, it were held that deity is the *next* emergent step after humanity, it might well be disputed whether such a conception was genuinely theistic. On such a theory there would be no reason why "God" should be anywhere near the top. He might only be the humble worshipper of his next immediate superior; and so on without end. Such a philosophy would be very doubtfully theistic.

Deistic theism may in various ways be a weakish and objectionable form of theistic theory. In other words an intelligent creator who was not a providence might seem to be too slick to be true.

Theism and Cosmology

It seems necessary, however, to examine the deistic part of theism rather closely, and the first series of these lectures will principally be occupied with that very enterprise, i.e. with what the deists asserted, postponing consideration of what they (perhaps weakly) denied. Deism *was* concerned with important aspects of the relation between cosmology and theism. It is proposed to discuss theistic (including deistic) questions, on a provisionally realistic basis, in the first series of these lectures. The second series will have the general title of "Mind and Deity" and will begin with an examination of the respects in which it may be claimed that the philosophy of theism would be strengthened if the provisional realism of the first series were corrected or supplanted by (it may be) a more adequate philosophy. In such a philosophy (it would often be said) the place of mind in the cosmos would be greatly enlarged. If so, God's moral and aesthetic attributes might also receive a new interpretation. For this and other reasons it seems advisable to postpone the consideration of divine providence to the second series of lectures. It will be the theme of the greater part of the second series.

III. *The Cosmological Argument*

The cosmological argument for theism is the argument from the self-insufficiency of the world. It has two parts, an existential assertion "the world exists" and an inference by complementary relation, "the world requires a divine complement". Because of the existential premiss the argument is quite different from, say, the Ontological Argument. The argument is very commonly used to-day, despite professed objections to traditional theistic demonstrations, and its general structure would seem to contain a very large part if not the whole of empirically grounded theism.

An apparently wider argument from "Something exists but is self-insufficient" is too weak for its purpose. If each finite thing were self-insufficient it would not follow that all finite things were together self-insufficient. Arguments of this type, therefore, are forced to proceed *via* the reputed self-insufficiency of the *mundus*.

The inference by complementary relations is next examined. Traditionally this inference is to the complementary of the "contin-

Synopsis of the Discussions

gency" of the world. Such "contingency" may mean (*a*) fortuitousness (*b*) logical or causal dependence (*c*) "the opposite of necessary".

The meaning of (*a*) must be that the world, which is not fortuitous (or chaotic), would be fortuitous (or chaotic) if it were left to itself. It is argued in detail that even if we could accept this hypothetical and very sophisticated premiss we could not say *what* the required complement was.

Regarding (*b*), the logical and causal senses of dependence are formally rather similar. The causal complement is said to be a "first cause". But if we could prove that the world is an effect, all that could be inferred would be that it must have *a* cause not itself, and not that such a cause must be a first cause.

Regarding (*c*) there is a serious logical oversight. There are no "necessary beings," because "necessity" is a modal predicate that applies to propositions only and never to things. In a wider sense all matter-of-fact is "brute" fact in the sense that, so far as we can see, it might conceivably have been otherwise.

In a still wider sense the self-insufficiency of the world may be said to be an expression for the world's incompleteness. This, however, if it were true, need not help us very much. If the world were a fragment, completion might be reached with another fragment. If the world's incompleteness is supposed to mean its inferior mode of being, the phrase is objectionable. What exists does exist and cannot have semi-existence. The fallacy seems due to the mistaken belief that there can be "epistemological objects". Again, if the world's incompleteness really means that it is illusory we should be denying, not asserting, the cosmological argument; for we should be denying its existential premiss. It would be wiser to try to dispel the illusion of a godless world by less pretentious and more direct methods.

The recent philosophy of "incomplete facts" seems itself to be very incompletely argued. On the other hand there would certainly appear to be intelligible philosophical reasons for holding that we may have no sufficient warrant for saying that we know there is a world (or a universe). "All that there is" might have no stronger status than e.g. "all the finite integers that there are". While extreme finitism in metaphysics seems untenable, a modified finitism might be all we can legitimately assert. If so, while we could attach a meaning to "everything that there is" we might not be able to speak intelligibly of "nature" or of the "world" as a single entity. Certainly, whether or not the world be infinite, we have no right to speak of it as if it were an ordinary finite thing like a table. This criticism, however, can scarcely be supposed to be helpful to the cosmological argument. That is the argument from the debility of the world at its proudest. If our right to talk about "the" world is challenged on finitist or semi-finitist grounds, our right to infer a God is challenged by the same logic. The complement could not be *another* (finite) thing.

IV. *Creation*

It is the developed, not the naïve, religions that emphatically assert the creation of the world. But the doctrine of creation, if true, would ensure that God could not be hampered from any external source (that is, if the doctrine of creation be allied with the doctrine of the possible annihilation of the world).

There are hesitations in the Book of Genesis, and a comparison of creation-theories with emanation-theories does not seem to be very instructive. There would seem, however, to be three clear logical possibilities: (1) that the world was made, but not out of anything (creation), (2) that it was made out of something (e.g. out of chaos), and (3) that it is some sort of output of divine substance (emanation).

The question may be raised whether there is creative process *within* the world, and whether such "creative" process throws any light upon the possible creation *of* the world. "Creative" imagination, creative biology and creative electricity are examined in turn. None of these seems to be creation *ex nihilo* and the sense in which any one of them *is* "creative" seems to be very obscure.

A much more important question for philosophical theology is whether creativeness is a general world-characteristic, and still more widely, a general characteristic of all existence, including a divine over-world if such an over-world there be.

A metaphysics of this type is suggested, resting in part upon Whitehead's analysis of the problem. Briefly stated, it is to the effect that the past at any moment is determinate but impotent, the

Synopsis of the Discussions

present determinate and potent, what we call the "future" indeterminate and therefore impotent. It is an assertion of the absoluteness of becoming.

In preparation for the next lecture (on "eternity") some indication is here attempted of the relation between such a metaphysics and the usual Christian theory of world-creation.

The theory of absolute becoming would allow that nothing simply pops into being, and consequently that the "world" does not. It asserts the inevitable continuity of each moment of existence with the preceding and with the succeeding moment. It holds in short that all continuance is a species of renewal (not necessarily a renewal of "the same"). Thus it resembles the Cartesian (and pre-Cartesian) theological doctrine of "continual (re-)creation". But it maintains that the fact is ultimate in all reality.

If we distinguished between "worldly" and over-worldly reality, the metaphysics of absolute becoming would apply to the over-world (or to God) as well as to the "world" (or under-world). It would admit the possibility that the underworld came into being at a particular moment, the overworld having previously existed from all eternity without any underworld, and that the underworld might at a later date be annihilated. What it could not admit would be that a world of process proceeded from a reality devoid of process, that a world of *genesis* was the "moving image" of a world of immutable *ousia*.

Since these matters are among the most difficult that human speculation has to face, over-confidence regarding them is all too easy and all too common. But they are not to be evaded.

V. *Eternity*

This lecture has three parts.

The first attempts to show the stubbornness and indeed the invincibility of change, time and the time-like in all reality. Even the illusion of change implies some change somewhere. Even "relative" change implies change *simpliciter*. In the simple time-like process of a dream there really is succession irrespective of all physical chronometry. There is also some apprehension of distance in succession and of rate of succession. Again, building upon the

Theism and Cosmology

results of the last lecture, we have to say that what is later than any moment is indeterminate at that moment. That is an absolute statement and cannot be affected by the relativity (to space, say) that may be implied in physical chronometry.

The metaphysicians who try to prove the illusoriness of succession do not seem to have succeeded. Their attempted proof derived from a supposed contradiction in (tenseless) assertions that express tense depends upon the neglect of tense and therefore fails. Their criticism of the durationlessness of "the present" is illogical, and the monster that some of them would put in its place, the so-called "specious present," should deceive no one.

The second part of the lecture deals with the possible meanings of "eternity". One such meaning is temporal everlastingness. That is the doctrine of everlasting continuants, a doctrine fully compatible with the view that all conservation is renewal. A second meaning is the "everlasting now". In this lecture it is argued, with special reference to mysticism, that there can be no *nunc stans*, no literally arrested passage. A third meaning is based upon the timelessness of truths and of general conceptions. Such timeless statements might be made about the flux and do not imply the possibility of timeless actuality, of timeless *things*. In general we need not hold that it was true that Caesar loved Cleopatra before he ever met Cleopatra.

The third instalment of the lecture deals with the relations between the first two instalments and theism.

It is maintained that changes in Y cannot wholly depend upon changes in X unless X itself changes in some way, that if X and Y are co-existent entities then, if one be temporal, the other must also be temporal, and that the only possible solution of the problem of change and permanence is the Heraclitean solution, viz. of stability *in* change not of stability *and* change. A Heraclitean persistence of pattern (not necessarily inconsistent with certain variations in the pattern) is conceivable, and there may be senses in which there is always a conservation of the preterite in every present moment that inherits from the past.

The indeterminateness of the future at any moment, however, must hold of God as of all else (if anything else there be). It cannot be true that for Him *futura iam facta sunt*. His determinate foreknowledge must therefore be denied. At any given moment there is nothing determinate for him to foreknow. Some discussion of

Synopsis of the Discussions

the relations between these matters and the theory of determinism is offered, with special reference not to determinism as a conjecture or policy in scientific investigation, but to determinism in metaphysics.

Finally, it is suggested that the conclusions of the third part of the lecture agree with certain very important facts. They accept becoming or process for what it seems to be. They admit the ultimate freshness of the present moment in our psychological experience as well as elsewhere. If they were true there would be no mystery in the fact that we cannot foreknow the future (but can only make plausible guesses regarding it) although we do know the present and do have retro-cognition (not conjectural) of the past in memory.

VI. *Ubiquity*

Theological thought in the West has wavered about divine ubiquity. On the one hand it was anxious to deny that there was anything God did not touch. On the other hand it was afraid that it would have to represent deity as being in some sense material if it represented him as spatial. In a more sophisticated way it might be maintained that spatial distance is aloofness, and so that divine spatiality would contradict divine unity.

In the first part of this lecture the phenomenology of space is discussed. It is maintained (1) that spaciousness is not wholly logical but is also and necessarily perceptual; (2) that not all our perceptions are directly and ostensibly spatial; (3) that if we learn of space by some sort of "intuition" (Alexander's term) such intuition is a sort of *construing* rather than a perceptible expanse; (4) that the unity of space is primarily an implication of the supposed unity of causal influence in physics; (5) that the traditional metaphysics of "space", its substantiality or quasi-substantiality, its continuity, infinity, homogeneousness and absoluteness, has often been more pretentious than the data warrant (it would seem, however, that there must be *something* absolute about spaciousness); (6) that the relation between space and the bodies that are said to occupy space (an important question when we are dealing with traditional materialism) carries us a long way beyond the sensible

Theism and Cosmology

appearances, and seems to resolve itself into the sort of topic discussed under the heading (4) *supra*.

In the second part of the lecture an attempt is made to consider some of the outstanding theological implications of this phenomenology of "space".

Since space is not merely logical both Sir James Jeans (with his God who for ever geometrizes by *pure thought*) and e.g. Malebranche with his doctrine of the intelligible extension in which we (intellectually) "see" all things in God cannot escape criticism. Some other traditional difficulties seem to be of a minor kind. If "space" e.g. were a "substance" there would seem to be no reason why God could not create or annihilate it as well as "body". On the other hand if it be true, as argued in (4) *supra*, that the unity of space is in the main an implicate of the supposed unity of all physical action, it is very hard to deny God's ubiquity if he be the supreme cosmic agent.

In modern philosophy it is not unusual to hold a mind-body theory according to which a spaceless mind directs and controls an extended body, and such theories are very often applied very confidently to the operations of a spaceless deity upon the cosmos. Some relatively modern theologians, however, have disputed the point, and the very vigorous opinions of Henry More in the seventeenth century would still seem to be pertinent.

According to More, God must either be nowhere, or somewhere or everywhere. The first suggestion (i.e. the "nullibism" of the Cartesians) was, he held, simply absurd. The second suggestion was timid and deservedly rejected as heterodox. The third therefore must be true. In other words God was literally ubiquitous. God, nevertheless, was a Spirit. Therefore his was an "immaterial or metaphysical" extension, a sort of fourth dimension in space. (More called this "divine spissitude" and suggested that it was a name for the coherence of the physical universe without which bodies would lapse into chaos. That was his final criticism of materialism, and Stout and some others say something similar to-day.)

More's axiom that "if a thing *be* at all it must be extended" does not seem to be self-evident. Indeed space is metaphysically less stubborn than time and might conceivably be illusory; and, as we have seen, "to be somewhere" is a most complicated phrase. Nevertheless the idea that mind or spirit is non-spatial in God or

Synopsis of the Discussions

in man is a hard thing and perhaps an unnecessary thing to believe. It may be as hard to believe as that life is non-spatial, an inference that cannot be drawn from the circumstance that life has no particular shape. If divine action is ubiquitous then God, in all probability, *is* ubiquitous in a sense that is not metaphorical; and that seems to be the conclusion that good theistic cosmologists should accept.

VII. *Omnipotence*

The idea of unlimited "omnipotence" may not be more intelligible than other attempts to dispense with all "limitations", and may lead to fantastic speculations about what a power literally *capable de tout* could do. For these and other reasons it seems better to discuss the *omnificence* rather than the *omnipotence* of deity.

The traditional argument is that natural causes are only "second" causes and imply a "first cause". (Nothing in this argument, however, compels the inference that there is just one "first cause".)

Such an argument seems based upon the belief that "first causes" are the only genuine causes—more accurately, the only genuine "cause-factors". It is necessary to examine the meaning of causality if we are to deal seriously with such a contention.

There are three principal views: the uniformitarian, the activist and the logical (or quasi-logical). It is convenient to discuss these separately although their interconnexions may be complicated.

Our first question is how the distinction between first and second causes should be interpreted in terms of each of these theories of cause.

On the uniformitarian view it is plain that there is no such distinction between first and second *uniformities*: and that, if the world is regarded as unique, there could be no uniformitarian theory of its origin. On the other hand experience might show sequences that had no uniform causes but proceeded uniformly once they had arisen, and it might show many non-uniform events ("uncaused") subsisting side by side with various uniform sequences.

On the activist view an active cause might be capricious, and it might be held that "passive" events borrowed all their appearance of efficacy from genuine active causes.

The logical or quasi-logical view might appear to be able to

Theism and Cosmology

hold that "first" causes have a status similar to "first" principles and "second" causes a status similar to subordinate principles. But causes and effects are always temporal, and in this reputed time-logic or quasi-logic it would seem that there is no room for such a distinction.

Our second question is the application of these conclusions to theology.

The uniformitarian view is not propitious to first causes, divine or not divine, although the doctrine would be consistent with a pantheistic interpretation of the uniformity of nature.

Much theology has relied upon an activist view of causality, having a special tenderness for the view that volitions are genuine first causes and that all other reputed causes are only passive echoes. This leads to difficulties. If human volitions are genuine first causes then, if men are not God, divine omnificence is denied. The idea that volitions are the *sole* genuine causes does not seem, however, to be capable of withstanding resolute criticism.

The logical or quasi-logical view of causality may be capable of meeting Hume's strictures by a flanking movement, and may attempt to meet the objections previously mentioned in this lecture by holding that what theology should assert is a First Ground rather than, strictly, a First Cause.

We conclude, however, that such a "ground" would not be a "cause" in any intelligible sense, and therefore hold that there is no clear road to a proof of divine omnificence on any one of the major theories of causality. While the failure of demonstration need not extinguish belief it should not be inferred that there must be some probability where demonstration is not to be had.

Some remarks are appended regarding the relations of the three principal theories of causality to one another.

VIII. *Teleology*

The Argument from Design may appear to be less abstruse than most other empirical approaches to theism, and some would say that there is no humbug about it. It survives to-day and may be introduced by a general examination of teleology. That is the subject of the present lecture.

Synopsis of the Discussions

The plainest meaning of teleology is idead conscious purpose. Such ideas or expectations act *a tergo* (although they look ahead) and should not involve any mystery. It is only dogmatic, not philosophical, to say that ideas cannot have any physical effects; and although value as such cannot operate, the idea or expectation of value or benefit may affect human agency.

Teleological action of this kind may include rather low-grade ideas, but there is a difficult problem regarding the point where the relevant ideas must be supposed to cease, for instance regarding the first performance of an instinctive action. It is not enough to talk vaguely about the presence of an "analogy", and say that we act "as if" we had ideas although, perhaps, we haven't any.

In idead teleology the ideas are (*a*) prospicient and (*b*) appear to be *sub specie boni*. In the first performance of an instinctive action, say, there is preparatory adjustment but there need not be any prospicience. "Value" in action *sub specie boni* is relatively seldom an explicit idea of "value" or benefit. The usual statement is that the projected action "seems good" to the agent. That is "value" in the very wide sense of "interest". Such "interest" may appear to be a natural bridge between idead and unidead teleology. The agent feels attracted but may not have any idea of good. It may reasonably be doubted, however, whether such feelings of attraction are either "ideas" or specific cause-factors in any action whatsoever. On the other hand it seems clear that there is unidead as well as idead teleology in nature.

Interested action touches the quick of our feelings, and many of our most vital biological impulses are felt in this way without being accompanied by any clear ideas about what is happening; but adaptation, co-adaptation and the like are patterns of natural action, are teleological patterns, and may be wholly unfelt and unidead.

Such teleological patterns seem plainly to be found in living creatures and there is no contradiction in unidead living. The view that biological teleology must somehow be "interested" seems to depend on the principle that teleology connotes value and that all intrinsic value must be mental. In this lecture it is asserted, on the contrary, that growth, reproduction and the like are maintenance-values which are not simply means, but are ends. Exploration of the analogy between idead and unidead teleology therefore yields

Theism and Cosmology

the result that unidead preparation should not be confused with idead prospicience and that unidead maintenance-values should not be confused with idead good consciously chosen as such or even with psycho-physical "interest".

Again, there seems to be no sufficient reason for holding that maintenance-values cannot be sub-biological or, generally, that adaptation, co-adaptation, using and being used may not pattern the inanimate. We should therefore conclude that unidead teleology is very prevalent in nature in the form of teleological *patterns* although not, as in idead teleology, in the form of teleological *cause-factors*.

The supposed contrast between machine-like and teleological is mistaken. Machines are of service and in that sense teleological. Most of them, it is true, have been designed by men, that is, they are the results of idead teleology. But they perform their services without ideas on their part, and there might be natural machines. If the question concerns being of service to man, berries and nuts perform that service without having been designed by man to do so. It is not self-evident that whatever is serviceable to anything must have been designed to be of service. Our own bodies are of service to our minds but it does not follow that they were designed by our minds or by any minds.

This conclusion is obviously relevant to the Argument from Design.

IX. *The Argument from Design*

In the traditional Argument from Design it was held that the particular adaptations in nature were far too numerous to have been unplanned, and also that the whole physical universe must have been planned.

The nerve of the argument was that adaptation bespeaks design, that apparently unidead teleology presupposes idead teleology. This inference was directly challenged in the eighth lecture. Teleological patterns, we there argued, may be quite unidead, directly or indirectly. Indeed all ideas empirically known seem to depend upon unidead life and not contrariwise.

In its most ambitious form, the Argument from Design holds that the universe is very like a machine and that all machines are

Synopsis of the Discussions

designed. This argument *might* be sound if its premisses could be established; for although arguments applying to the parts need not apply to the whole they *may* so apply. But is the universe like a designed machine, like a watch, say? Again, since something must be ultimate, what reason have we for denying that the material universe may be ultimate? Why should the fact, if fact it be, that it is orderly and patterned be irreconcilable with its ultimacy? Or, once again; much that is most highly organized within the universe (e.g. living organisms) appears to arise in the unidead not in the idead teleological way. If organisms are not machines, much in the universe is not machine-like.

Leaving these generalities, let us ask whether there are special reasons for God-like design in or of the universe?

What designs would we attribute to deity?

It may be suggested that God for his own glory designed the vastness, the order and the beauty of nature; but these arguments seem to be rather weak. We have no means of judging how glorious the size of nature is; for size is relative. While many pious minds, especially in the seventeenth century, have come near to identifying orderliness with divinity, and (with some hesitations about miracles) have asserted that natural law is infinitely superior to providential piece-work, their arguments are unconvincing. They seem to be talking about laws sufficiently simple for the human intellect to manage. As to beauty, the arguments in favour of the view that what men call beauty is relative to *human* percipience are too strong to support an inference to absolute, divine or super-human beauty, and there is no force in the contention that natural beauty presupposes either a natural or a non-natural artist.

Other suggestions of this kind refer principally to divine providence and are here raised briefly, although the proper place for them is the second course of lectures.

One of the chief may be described as a problem in moral ecology. Is the earth's crust peculiarly adapted to the human intellect, to human or animal happiness or to the formation of moral character?

We are not in a position to say how good a laboratory the world is, although it is not unpropitious to human science. Regarding human and animal happiness we may reasonably believe in a temperate optimism and in a guarded meliorism, but have insufficient natural data for deciding whether the fact requires a super-

Theism and Cosmology

natural explanation. There may be interminable argument on the question how good a moral educator the natural universe is (utilitarians and deontologists, for instance, having different standards). In short, moral ecology rests on an indeterminate basis both regarding men and animals and regarding the presumed inhabitants of other planets.

Similarly, if instead of considering the fitness of the environment to the users we consider the origin of the users themselves it is at least possible to argue that the users (e.g. men or animals) arise in the unidead and not in the idead teleological way.

The Argument from Design seems indeed to be weak; far weaker than arguments to teleology (perhaps deiform) that is *not* design.

X. *Examination of Cosmological Theism*

A general and rather free statement of the course of this first series of lectures is here attempted.

Theism must at the least be cosmological, and cosmological theism argues from the double premiss that there is a world and that the world is self-insufficient. Even the Argument from Design is of this general order, for its assertion is that the world must have been made by a designing Mind.

Theism cannot be *demonstrated* on cosmological grounds, and the failure of demonstration is not, as such, a positive encouragement to theists, although the possibility remains that what cannot be demonstrated may nevertheless be reasonable to believe.

Speaking generally, cosmological arguments for theism are apt to be the weaker in proportion as they attempt to prove the existence of a quite over-worldly deity. Their tendency, indeed, is towards pantheism. One reason for this is that time and change are metaphysically inexpugnable. Space may be less stubborn. It *might* be illusory; but cosmologists do not usually argue from the *illusion* of the cosmos.

As remarked above, cosmological arguments for theism are based (*a*) upon the world's existence, (*b*) upon its self-insufficiency.

As regards (*a*), it is a fair and a difficult question whether and in what sense we are entitled to say that there is a world, nature, the

Synopsis of the Discussions

(physical?) universe; but the more extreme assertions of philosophical empiricists on this question, including the most up-to-date ones, need not be accepted.

As regards (*b*), the essential question is whether the world's self-insufficiency can be proved, and, if it can be proved, whether it can also be shown that the required complement must just be God.

One of the most usual arguments is cosmogonical. The world must have a Cause, it is said, and that cause or source must be God.

Such a cosmogony might be argued on scientific lines, e.g. that the galactic universe must end in heat-death and that anything that has an end must have a beginning. But it is not self-evident that having an end and having a beginning have this or any other logical correlation; an end in heat-death is not annihilation; and the entire scientific argument is speculative and also doubtful in many ways.

The presupposition of the argument from self-insufficiency is that only the self-sufficient can exist. This may covertly involve ideas about the sufficiency or "perfection" of existence that may reasonably be denied.

It is not convincing to hold that the world is self-insufficient because illusory on the ground that it is infinite, fragmentary or an effect. If the world be not illusory, it is hard to show that it could not be ultimate. Whatever exists really does exist. It cannot have an inferior non-ultimate "mode of being". If it were "brute fact" it might ultimately *be* brute fact.

We should not assume that if the world exists it alone can exist, with the consequence that deity, if there be deity, must be just the deiformity of the world. But we should not fly to the opposite extreme, and, demanding a "perfection" that we do not find in the world, assert that a supra-mundane "God" therefore exists.

As before remarked, cosmological theism tends towards an immanent rather than towards a transcendent theism. Hence it takes pantheism very seriously. Pantheism need not "take the easy course of finding God in things just as they stand".

I
Concerning Natural Theology

In the lectures that are to follow I intend to talk about the subject I was asked to talk about, namely natural theology. I find the subject fascinating, and I believe it to be very important. Therefore, although I am profoundly diffident about my own ability to contribute usefully to the discussion of so great a theme, I am drawn and tempted into that very enterprise.

Two preliminary objections stare into my eyes. The first is that the distinction between natural and non-natural theology is often supposed to be altogether outmoded in the times in which we live. The second is that the adjective "natural" may be thought to be too ambiguous to be of service to any boundary commission whatsoever.

The first of these objections does not seem to me to be very serious. Fashions change, always from causes, not always for sufficient reasons. The second objection, however, is very menacing indeed. I do not think that any answer to it is wholly satisfactory, although I shall try to show that we may legitimately attempt, despite a certain imprecision, to develop a "natural" theme by "natural" methods.

I allow, however, that the phrase "natural theology" is not above suspicion, and has to be detained for questioning. So to the question.

There have been two principal ways in which "natural" theology has been contrasted with theology that, in some important sense, is not "natural". The first has descended from antiquity, and is the distinction between philosophical and unphilosophical theology. The second, of later origin although now more familiar in the West, is the distinction between a theology that is not based upon "revelation" in some special and peculiarly significant sense of that term, and a theology that is so based.

Let us begin by considering the older division.

M. Terentius Varro, Cicero's contemporary, drew a threefold distinction between natural, mythological and civil theology, and his influence endured. (Tertullian and Augustine quoted him extensively. Montaigne, for instance, many centuries later, called him a "notable divine".) Indeed it became quite usual to regard metaphysics and natural theology as identical. Thus Leibniz wrote as follows to the Landgraf Ernst von Hessen Rheinfels: "Since what perfects our mind (if we leave on one side the illumination of grace) is the demonstrative knowledge of the greatest truths by their causes or reasons, it must be admitted that metaphysics or natural theology, which treats of immaterial substances, and particularly of God and the soul, is the most important of all."[1]

Let us, then, examine Varro's distinction between the "natural" (or philosophical) and the two other species of theology.

To Varro and his circle all mythology was but a fable of the poets. Modern anthropologists, no doubt, would dispute this opinion, but would not have much to say about it that would tend to bring mythology within the ambit of philosophical theology. Man, it is true, may be a mythologizing animal, and the greater theological myths may yield important evidence regarding what, after Hartley, we may call man's "theopathic faculty". But mythology is not philosophy. Again, the constancy and the recurrence of some of the larger human myths in widely different epochs and places may yield a legitimate domain for the application of the famous Vincentian test "Quod ubique, quod semper, quod ab omnibus creditum est". Fabulation may be a deeper and even a better thing than much that we call rationalization. It may have

[1] *Philosophical Writings* (Everyman), p. 235.

Concerning Natural Theology

"lyrical validity". But it is not science and it is not philosophical theology. Moreover it is liker poetry than either of these.

Certainly we must all allow that the boundaries between what is and what is not poetry are also very hard to draw. If poetry could be defined by its theme, science, philosophy and philosophical theology might well be among its proper subjects. Lucretius is the greatest witness, and Dryden, as Mr. Michael Roberts has recently told us,[1] said that "a man should be learned in several sciences, and should have a reasonable, philosophical and in some measure a mathematical head, to be a complete and excellent poet". That is a possible view, and the instance of Lucretius shows, as is evident in any case, that the presence or absence of metrical form could not supply an accurate ground of distinction. In short, philosophy or theology might be treated poetically, and some have held that the greatest truths and the deepest surmises about the nature of things should all be in poetry. They may think, like Whitehead, that what is deep must always be dim.[2] Again, they may find themselves in agreement with the late Mr. W. B. Yeats: "I thought that whatever of philosophy has been made poetry is alone permanent, and that we should begin to arrange it in some regular order, rejecting nothing as the make-believe of the poets. I thought . . . that if a powerful and benevolent spirit has shaped the destiny of this world, we can better discover that destiny from the words that have gathered up the heart's desire of the world, than from historical records, or from speculation, wherein the heart withers."[3] On the other hand, if philosophy restricts itself to reasoned argument, and is, in that sense, scientific in its conception, poetry, on any ordinary interpretation, is neither its ethos nor its appropriate medium. In this sense, philosophical theology has to be distinguished both from mythopoeic and from poetical theology.

To say this is not to say or even to hint that philosophical

[1] *The Modern Mind*, pp. 95 f.
[2] In *Modes of Thought*.
[3] *Collected Works*, VI, 71.

Theism and Cosmology

theism is truer, or deeper or better than poetical theism, or that the two must pull different ways. Indeed, Montaigne condemned them both on the same grounds. "Truly", he said, "philosophy is nothing else but a sophisticated poesy. Philosophy presenteth unto us, not that which is or she believeth, but what she inventeth as having most appearance, likelihood or comeliness." Even in that sceptical statement, however, it was admitted that the philosophical kind of vanity had a certain kind of argumentative sophistication without which it would miss its apparent effect. The difference is equally clear even when poetry claims to have truth for its goal no less than philosophy or science. Wordsworth said that poetry was "the impassioned expression which is in the countenance of all science", thus indicating, among other things, that science and philosophy may conceal the personal emotions of fervent truth-seekers while poetry glows with these personal emotions. Regarded as truth-getters, however, poetry and philosophy are not the same.

If, in the fashion of the day, we say of language what used to be said of thought, we might remark that the language of poetry attempts to convey an atmosphere, and that the language of science and of scientific philosophy is meant to be taken literally. In another of the happy phrases Mr. Roberts has culled from the past, philosophy, we may say, should aim at a "close, naked, natural way of speaking". I should prefer, however, to use the older idiom and to say that the distinguishing mark of philosophy and of science is to look for the reasons of things, and to set these down faithfully in order to show their explanatory value. No doubt actual philosophy and much actual science do not entirely restrict themselves to this policy. They are often, as we say, less grimly inhuman. Nevertheless the giving of reasons, and the orderly procedure by reasoned argument, every step being clearly shown, is the essence of all *scientia*, and is not poetical at all. A poet may lisp in numbers, but he does not even stammer in sorites.

In their several ways both *scientia* and poesy may enlarge

the imagination, liberate the soul, and transfigure the world, but I shall not stay to enquire minutely into the difference between scientific and poetic imagination, between scientific and poetic disimprisonment from vulgar prejudice, between the attitude of a philosophical "cosmotheorus" and of a poetic spectator of all existence. I shall also not enquire here into the motives that induce men to turn to poetry, or to science, or to philosophy. These motives are varied. Men may study science for economic purposes, for patriotic purposes, for propagandist purposes. They may study philosophy because they think it is the palladium of truth, or because they think the subject has a certain distinction and pleasing rarity. They may turn to poetry for beauty of cadence, for luxury of moving imagery, for the play of verbal dexterity, for release from all that is drab, or hoping rather wistfully for a mystical narcotic. In short they may study all these things from causes that tell us very little about what the things themselves are. The distinction between reasoned and unreasoned, however, is very much tougher. In science and in philosophy, we may say, an essential requirement is that the books should invite and be ready for audit. In poetry there is no such invitation and there are no preparations to meet it.

As I have indicated, I am insisting so resolutely upon the conjunction of science with philosophy, and on the disjunction of both of them from poetry, largely because the contrary view is so often expressed to-day. According to this modish body of opinion (as according to Montaigne), the greater part of traditional philosophy, and the whole of philosophical theology, should be regarded as a kind of poetry, appealing to the heart or to the fancy but, to a mature and critical mind, not very much more than a fairy story. Such statements may be made in quite a friendly spirit, but are usually meant to be derogatory. Their purport is that the salt of traditional philosophy was the science in it, and that the scientific part of it alone is worth saving. "Speculative" or spectacular philosophy, we are told, is little better than a well-fabulated myth.

Theism and Cosmology

According to this view a clear-headed modern philosopher should not try to erect cosmologies, to build theologies, to interpret theophanies, or even, like Marx, to construct a semi-economic philosophical fable. In the past (we are told) such attempts acquired an unearned increment of unwarranted prestige because of the presumed alliance between this kind of philosophy and the scientific part of philosophy. Nowadays (it is said) that support must be withdrawn and most that is called philosophy is only a species of poetry, a species moreover which, regarded as poetry, is not the best in that kind.

When Hume,[1] like a second Caliph Omar, proposed that our libraries should be cleansed and also greatly lightened by committing to the flames all the volumes that did not deal either with formal mathematical demonstration or with experiments into matter-of-fact, it was the books about divinity and school metaphysics that he thought most deserving of combustion. The logistical positivists of to-day say the same thing with all the emphasis at their command. Kant's "critique of all theology" was not wholly dissimilar, and Comte, as everyone knows, held that the metaphysical as well as the theological stage of speculation had to be outgrown before the human intellect attained its true majority in the understanding of positive science. Such a contemporary writer as Neurath[2] follows Comte very closely indeed.

In opposition to these prevalent views, then, I am asserting that philosophy, philosophical theology and metaphysics, as I interpret these studies, are not a kind of poetry, and should not be confused with poetry. I do not say that they are sciences in every sense of that hospitable term. They may not all be of the purest analytical water or meticulously experimental, but they are on the side of *scientia* in the general sense that they profess to be based upon reasoned argument, and, as I have said, to have their books ready for audit. How much is claimed for such enquiries is another story. "I have never

[1] At the end of his *Enquiry Concerning Human Understanding*.
[2] O. Neurath, *Empirische Soziologie* (1931).

Concerning Natural Theology

attempted to declare 'the' truth" said Pherecydes in his "letter" to Thales.¹ "I have simply said what a man may say who speaks of the gods. The rest is only conjecture." That is still the spirit of natural theology. We are still asking whether a reasoned enquiry into the nature of things is evidence of the deiformity of reality. We still want to know whether we can only weave poetical myths about such matters, or can build on a solid foundation of rational argument.

Varro's other contrast, that is to say the distinction between natural and civil theology, may here be considered very briefly.² It would be a very different thing, it is true, if we were discussing religion and not theology. We could not then be brief; for although there is a sense in which a solitary man may find and keep his religion, religion, in the main, is a social thing whose finest flower is a spiritual community. Such a community, of course, need not be identical with any political body. The Church Catholic (although it does not as yet extend over all the world) is larger, even to-day, than most swollen empires, and Churches (to keep to Christian communities) have been persecuted and also tolerated where they were not nationally established.

Still, religion lays claim to a man's whole life, and political government claims the right to conscribe a man's body as well as to tax his goods. Therefore, even if we distinguish in the modern way between community, on the one hand, and political and ecclesiastical government upon the other, it is clear that all three are too closely conjoined to be dissevered without violence even in a notional way. Germany's recent embarrassments in this matter have not been arbitrarily manufactured. One remembers Hobbes's view that private judgment in matters of religion was the principal cause of

¹ Diogenes Laertius, Book I.
² It was an important contrast to Varro: "If he were founding Rome", he said, "he would have consecrated the gods and their names in accordance with the scheme of nature. But as it was, the State being long established, he wrote with the purpose that the masses might be willing to worship their gods rather than to despise them." A. D. Nock's statement in *Conversion*, p. 225.

Theism and Cosmology

sedition in the Europe of his time. We may also recall Hegel's doctrine[1] that the state is the wisdom of the world and its freedom the only earthly freedom, with the consequence (as he thought) that the relation of worldly to divine wisdom and freedom is philosophically as well as practically of paramount importance. In a minor way we may remember Lord Melbourne's *mot* "Things have come to a pretty pass when religion is allowed to invade private life".

Nevertheless such affairs have little bearing upon natural theology in its usual historical sense. Communal metaphysics (*pace* Hobbes and the Bolshevists) is as plain an ineptitude as communal physics would be. The same is true of communal theology unless it can be shown by general argument that some church or nation is, in a peculiar sense, God's peculiar people.

Varro's "civil theology" in the main concerned ceremonial religion. As such it was quite distinct from natural or philosophical theology; as anyone, nowadays, can perceive. Indeed in the later development of our tradition, it was usual to maintain that natural reason could not treat of the sacred at all. Medical and sanitary rules (like those in the Pentateuch) should be justified by medical and not by theological arguments—or so it was said by the advocates of natural reason.

This, however, introduces the second major contrast that I have mentioned at the beginning of this lecture, the contrast, that is to say, between natural theology and a theology that has revelation for part of its basis, or, as some contemporary Protestants frankly assert, the contrast between philosophical and Biblical theology.

Everything that we know, whether in whole or in part, may be said to be in a manner revealed, and all such revelation, for a Christian, must ultimately have its source in God. That being understood, however, there is plainly a distinction, in the Christian and in some other theologies, between the sort of theological proposition that may be collected from the

[1] *Philosophy of Religion*, English translation, I, 247.

Concerning Natural Theology

Book of Nature, "that beautiful frontispiece to eternity" as Traherne called it, and the sort of theological proposition that cannot be grounded on philosophical or scientific evidence but is based on the authority of the Book of Life. In Christian theology (if I may speak in the large) this distinction became the distinction between the comprehensible and the mysterious part of the Truth, the Truth itself being something that God had somehow made manifest in Nature, but had also declared in the Scriptures, and had infused into a saintly tradition through the gradual working of his Holy Spirit. A Christian receives the Truth as well as the Way and the Life, but still has to ponder the question whether on the one hand he has to receive the Truth with reverent mental vacuity, or as a *kerygma* of history divinely intended for the perpetual instruction of mankind, or on the other hand, very largely, as something that may be ascertained by philosophical and scientific methods.

Hence, from the nature of the case, there is and has always been a schism among Christian theologians between those who welcomed and those who despised or feared an independent (if pedestrian) scientific or philosophical investigation into matters the answer to which was already known beyond any peradventure. The one party argued that if Christianity became philosophical and made its peace with the culture and the science of the world, it would be all the stronger and the better equipped for presenting a victorious ideology. In antiquity the contention was that pagan philosophers such as Plato had offered proofs of theism, and that the atheists had been worsted in argument, as well as the Epicureans who, although they were not atheists, had denied God's providence. The better pagan philosophy, in short, was said to have been favourable towards theism. Therefore it was held to be unreasonable to expect that Christians of culture and of intelligence should deny themselves the advantages of this favourable atmosphere. Why should anyone stubbornly assert that pagan support was more dangerous than pagan opposition? The other party argued, with obvious force, precisely that this danger was very great indeed. Philosophical theology, such men said, was a theological hindrance and also a theological menace. It was a hindrance because it was a useless encumbrance. What was the sense of straining one's wits to prove a foregone conclusion? It was dangerous, because if the proofs of philosophical theology were weak in any respect, doubts concerning the Godhead were almost certain to creep in. That was the penalty of choosing weak ground where stronger ground could be occupied. When strength and audacity are yours for the asking, why should a bland but timid policy be allowed to prevail? Theological "appeasement" and theological disarmament are equally dangerous.

It seems clear that the early decision of educated Christians to think philosophically about God, and to include a philosophical theology in their world-view was greatly (although perhaps not unequivocally) to the advantage of Western culture. It may not, however, have been the best thing for *Christianity*, and St. Paul, in one of his moods, may have been wise when he told the Colossians[1] to "beware lest any man spoil you through philosophy and vain deceit, after the tradition of men, after the rudiments of the world, and not after Christ". Mr. Barth and his followers say the same sort of thing to-day, and although we are often told in this year of grace that the lion of modern science has lain down most amicably with the theological lamb, there is still a fairly general disposition among theologians to look askance at philosophy in its modern academic sense. Many indeed would be prepared to say that the attempt to make theology philosophical had been a disaster for the Christian Church. Philosophy, they would say, has always been a mutinous handmaid of the faith, a sort of Hagar, and so should be treated like that unfortunate lady. It is a fatal error to exaggerate the importance of philosophy and of philosophical technique

---

[1] Col. ii. 8. Cf. Bacon to James I: "The vain and indign comprehensions of heresy and degenerate philosophy".

Concerning Natural Theology

in the things of the spirit. The knowledge of God should not be transformed, in any way at all, into a hard, technical, hair-splitting metaphysical theology exhausting itself in dreary profundities. The schools of divinity should never be allowed to become an intellectual gymnasium that, religiously speaking, trained all the wrong muscles.

It seems abundantly plain that much in the Christian faith could not conceivably be reached by general philosophical methods. We need not here enquire how far the doctrine of the Trinity, say, is actually contained in the Scriptures, and how far its mystery is a piece of non-natural metaphysics subtly devised by sundry Church Councils. It is enough to remember that Christian doctrine could not be what it is if Christ's incarnation in Palestine and his resurrection are not to be regarded as historical fact. For Christians there *was* such a history, quite literally and not as a piece of *Urgeschichte*,[1] a sort of eternal essence of occurrences that never occurred. Nevertheless the *historical* biography of Christ, in the sense in which it has to be accepted in Christian doctrine, is as non-natural as any myth.

As it seems to me, it is also misleading to say, like the Archbishop of York in his recent lectures here in Glasgow, that the Bible (and its progressive clarification by the work of the Holy Spirit) pertains to natural theology because it is a record of religious experience. The Bible is certainly a religious document, that is, it is evidence of man's theopathic state of mind, but it cannot, on that account alone, be regarded as a philosophical premiss or even as a philosophical commentary. If anyone says that the Bible bears the stamp of truth upon its countenance, he has to be told that philosophers, in the exercise of their profession, cannot rely upon those impressionistic methods. (I must here ask to be excused from considering what kind of certified fact the Bible displays after historians and textual critics have finished with it.)

"Natural" theology, therefore, could never be the whole

[1] As for Brunner, *The Philosophy of Religion*, chap. 7.

Theism and Cosmology

of Christian theology, even if it uniformly supported that theology. It must always be less than Christian, and from the little that I know of non-Christian theology would seem to be narrower in principle than the theologies of many other religions. That in itself is not an objection to natural theology; but there would be serious reason for complaint if the limitations of the sphere of natural theology were sedulously and not quite candidly concealed, if, as J. S. Mill said three generations ago, "the whole of the prevalent metaphysics of the present century is one tissue of suborned evidence in favour of religion".[1] Some people seem to think that it is the function of Gifford lecturers to perform this very office. I repudiate that idea. If it were so I cannot think that these hirelings would be just what they have been.

But enough of that. If the true theology, or (for that matter) if Christian theology has to be regarded as a supernal manifesto that can neither be questioned nor pondered but has simply to be accepted in all its shining effulgence, many, I think, would experience a deep and wistful regret. This attitude, if it were inevitable, would imply that there was simply no sense in thinking about these high matters in genuine candour. The one thing needful would be docility, and a thoughtful docility would be little better than a thoughtless. For the diminution of mystery, if it could be accomplished, could scarcely be solid gain. Mystery, by hypothesis, would always be at the top.

These questions, however, do not concern us here. The Christian's best policy may be to renounce natural theology and all its works. The idea that philosophy should take God in hand—if Christian opponents put the matter so—may well appear to him to be as sacrilegious and absurd as the youthful Renan's famous remark "La science organisera Dieu". (Kierkegaard thought that a philosophizing theology was just a harlot: "Theology sits by her window in powder and paint, courting the favour of philosophy and selling her charms.")[2]

[1] *Three Essays*, p. 72.   [2] *Fear and Trembling*, English translation, p. 38.

Concerning Natural Theology

We, however, have to deal with these affairs not as Christians, not as Buddhists, not as Mahommedans, but as philosophers; and unless philosophy is itself mere vanity—a point I am not going to argue now—there is no doubt at all that it is a philosopher's proper business to treat among other things of theology or, as it was sometimes called, of "first philosophy". If theism is only a hypothesis, it is, as we shall see, a major philosophical hypothesis. If it, or any part of it, is more than a hypothesis and is demonstrable, that also is properly and indeed exclusively a philosophical theme. I need not cite authority, even of a philosophical kind, although I am not afraid of ideas that seem to be too good to be new. Those who believe that there is and has been for millennia only one great philosophy, called in the Latin "Philosophia perennis", would be the first to deny that *this* perduring philosophy has been wrong (and perhaps naïvely Leibnizian, if any Leibnizian can be called naïve) in holding that God, freedom and immortality were the eminent domain of metaphysics. These topics, however, *are* metaphysical even if they are also the themes of myth, poetry and religion. That is our ultimate, our only and our sufficient justification for dealing with them here.

In short, there is such a thing as philosophical theism, and "natural theology" is another name for it. Such a theism claims the liberty that is essential to all philosophy. It has the privilege of following the argument whithersoever it may lead. It is critical of outside authority, not under subjection to such authority. From that point of view, it does not greatly matter whether the reputed authority is natural or supernatural, civil or churchly. "Can you think at all", Blake asked in one of his moods, "and not pronounce heartily that to labour in knowledge is to build Jerusalem, and to despise knowledge is to despise Jerusalem and her builders?"

That, I think, is all that need here be said about the contrast between a merely philosophical theism, and a theism that is based, in part or entirely, upon the authority of some special revelation. I may be told, however, that the term "natural theology" in its historical connotation has come to imply a good deal more than the relatively colourless term "philosophical theology". Has not "natural theology" been commonly taken to mean the theology that was legible in the Book of Nature, or, alternatively, as the truth about God that could be read by the light of Nature?[1] Has not the term been drenched with these and with similar associations? Is it not reasonable to ask whether such associations are *only* associations, elf-horns faintly blowing in an archaic haze, or whether they have still an intelligible and defensible meaning? And if they are *only* associations may they not now be indelible, as well as mazily intricate? Might not that be true even if Bacon was right when he called natural theology "that rudiment of knowledge concerning God obtained by a contemplation of his creatures"?[2]

That, as I said at the beginning of this lecture, is one of the principal questions that we have to discuss. It, more than any other cause, compelled us to detain "natural theology" for questioning, and I shall employ the rest of this lecture in the examination of it. I do not expect a pellucid answer, but I may possibly achieve a certain advance in clarity. Alternatively I should plead that although the theme is threadbare, brittle and dry, it cannot be avoided in any serious discussion.

It is beyond dispute that the word "nature" has been and still is used in a thoroughly ambiguous way. (1) Sometimes it means *all* that there is, and in that sense nothing can be non-natural. (2) Sometimes it designates what is called the *cursus ordinarius* of events. In that case all that is radically exceptional and egregiously atypical, as well as all that either is or is called miraculous, diabolical, sub- or super-normal would have to be excluded from "nature". (3) In a third sense the "nature" of anything may be said to be just its constituents. The nature of water is H_2O. If so, the history of the water when it is drunk or spilt is not a part of its "nature". (4) Again, on

---

[1] "Pagan darkness" according to opponents.
[2] In *The Advancement of Learning*.

Concerning Natural Theology

the assumption that the really fundamental question about anything is "Where does it come from?", it is usual to say that the primitive and primordial is pre-eminently "natural". (5) Another favourite distinction is between the natural and the artificial. In terms of this distinction it may be natural for human beings to be cleanly, but it would be unnatural to use soap or a toothbrush. And there are a great many other ways of distinguishing between natural and unnatural. Indeed I have seen a list of such meanings that resembled the scoring sheet of a long innings at cricket, and I concluded that the batsman was still not out. (His precise score was 61.)

The five senses of "natural" that I have enumerated are, however, the most important, and they differ greatly in the degree of their relevance to our present subject. The first meaning and the fifth seem to be altogether irrelevant. The first, since it includes everything with equal avidity, excludes nothing and affords no basis for discrimination. The fifth is irrelevant from the nature of the affair. Science, philosophy and theology are all artificial in the sense that they require technical skill of a high order for their efficient development. Naïve theism is of no greater account than naïve economics or naïve trigonometry.

The third and fourth meanings yield more scope for discussion. As regards the third, the candle-power of the light of nature does not seem to be easily resolvable into separate constituent units, although the Book of Nature is often supposed to be written either in an ultimately mathematical alphabet, or in a more extensive but still in a limited set of mathematical symbols. As regards the fourth, it is clear that the origin of things is one of the traditional problems of natural theology, but if this problem is taken to mean the non-natural origin of "natural" things, it entails quite another sense of "nature" and of "natural".

That sense, substantially, is the second on our list, according to which "nature" is the "ordinary course of things" and the "natural light" is man's ordinary powers in the way of perceiving and of reasoning. In this sense of "natural" it would be unnatural if unusual events did not sometimes occur, but the theory, without any portentous strain, can adapt itself to such a situation. It speaks in terms of the "natural" *order*. Its theme is the *cursus ordinarius* of events. This order includes infrequent occurrences.[1] Similarly, with respect to man's "ordinary" powers, it may be contended, very intelligibly, that such powers may be exceptionally keen. The natural order of human intelligence includes the minds of very clever people who have cultivated the "ordinary" type of intelligence to an unusual pitch of perfection, or have inherited a quicker and a subtler wit than most. What has to be banished from "nature" in this sense, therefore, is the events (if there are any) that are opposed in their origin or in their character to the *order* of the *cursus ordinarius* of events. What is excluded is anything that is definitely super-normal, super-natural, or miraculous. "Nature" is just Physis and excludes all that Rabelais called "Anti-physis".

In this sense of "nature", "natural theology" would be the theology that is legible in the ordinary course of events, or, alternatively, the theology that men can reach by the use of their "ordinary" powers. These two conceptions need not coincide. It would not be absurd to hold that the ordinary course of things is, except superficially, beyond man's ordinary powers of comprehension. For the most part, however, it is commonly supposed that there is a very close connection indeed between the two. Indeed it may be suggested that there is a rather flagrant circle in the theory. Man's "ordinary" powers are those that are capable of assimilating "ordinary" events, and "ordinary" events are those that are assimilable by man's "ordinary" powers. The understanding determines nature and nature determines the understanding.

I shall try, firstly, to examine and to meet a general objection to this conception of "nature" and of "natural". Secondly, I shall attempt a more detailed investigation.

[1] I.e. it is usual to have *some* surprises. Consider football pools.

Concerning Natural Theology

The most general objection is that "natural theology" in its traditional meaning is a piece of grand metaphysical larceny. The Book of Nature (so-called) is said to be only a book of selections, a worldly-minded anthology. According to such objectors the naturalizing or secular anthologists leave God out, and so convey the impression that there is nothing divine in the largest book of all, the Book of Reality. The larceny itself may be described in various terms. A Christian philosopher, that is to say, someone like Malebranche who was much more than a philosopher who was also a Christian, would say that "nature" is a pagan effigy. It is something undeified, unsanctified, dismembered into a specious finitude, an unintelligible procession of "second" causes, although "first" causes are the *causae causantes* of all life and of all power. Hegel, with a different purpose—for he held that religious doctrine cannot be more than an imaginative sketch of the truth that philosophy alone can seize—made a very similar complaint. Nature, for him, was but an appearance, an Other that was manifested to mind, an Other that seemed to be self-subsistent but was really dependent, an Other whose truth was concealed until it was caught up into logic and united with rational thought. It may not be entirely plain what Hegel took this Other to be. Sometimes he described it as sensuous immediacy, sometimes as the world of finite things, sometimes as mere substance when the truth was Absolute Subject. But the general trend of his argument is sufficiently clear. "Nature", he held, was only a myth—a laboratory story if not precisely a fairy story—and the acceptance of *such* a "nature" was little better than idolatry. It betrayed hooded eyes and a vacant head. As Milton said, "Their eyes how opened, and their minds, How darkened".[1]

Clearly these strictures have to be faced. "I wish the lecturers to treat their subject as a strictly natural science," Lord Gifford said in his will, "the greatest of all possible sciences, indeed, in one sense the only science, that of Infinite

---

[1] *Paradise Lost*, IX, 1053–4.

Theism and Cosmology

Being, without reference to or reliance upon any supposed exceptional or so-called miraculous revelation. I wish it considered just as astronomy or chemistry is." I believe I can obey the spirit of these injunctions, and, in the negative part of them, am glad to obey *au pied de la lettre*, but it seems to me to be necessary to concede that theology cannot be treated *just* as astronomy or chemistry or biology is treated. The totality of things, if there be such a totality, is not one thing among others, and God, according to some theologies, is the *omnitudo realitatis*. The birth of a world cannot readily be compared with the birth of a child, and a world-ground cannot be a "cause" in precisely the sense in which a spark may be called the cause of some huge explosion. Juries are told, very properly, to employ their intelligence in the same way as they use it in the common affairs of life. The standard of proof in their case is not affected by the fact that the life or liberty of a fellow-subject is an affair of more than ordinary gravity. Similarly the gravity of theological issues should not of itself affect the type of evidence that is appropriate to them. The gravity of them may prescribe greater carefulness, but it need not prescribe more. In theology, however, the problem itself may demand somewhat different methods. If deity were a ninety-third chemical element, chemists should deal with him as they deal with the other ninety-two; but if he is not (or is not simply) one being among others, the nature of theistic argument may be affected in a logically relevant way.

The substance of the general criticism we are now considering is that "nature" is not reality but an impoverished aspect of reality, that reality suffers violence when it is treated simply as "nature", and that natural theology, because it offers this violence, should pay the penalty of its godless temerity. "Nature" in the plain mouths of to-day is reality denatured and desecrated. Such an objection is at least plausible and so deserves a nearer inspection. It may be considered in connection with the other point I have to discuss, i.e. man's "ordinary" powers in the way of knowing, or, in other words, the co-called

Concerning Natural Theology

"light of nature", together with the propositions and the realities that are said to be known by this natural illumination.

Very often the "light of nature" was declared to be just the "light of reason", and "reason" was regarded as the "candle of the Lord", distinguishable indeed from the stronger illuminant of grace, but sufficient for human science, philosophy and practical sagacity. Without this candle a man would be feeble-minded. With it he has all logic to guide him. It was a candle beyond all price generously bestowed upon everyone, Christian or heathen, provided he was not an imbecile. As Benjamin Whichcote said, "Things themselves speak to us, and offer notions to our mind, and this is the voice of God".

This general conception may need to be clarified. Many of its exponents may have exaggerated the function of "pure" or of "mere" reason. Indeed "natural religion" was understood by some of our eighteenth-century writers (e.g. by Clarke or Wollaston) to be pure rational morality, as cold, serene and indisputable as arithmetic. It may therefore have been proper to complain that pure reason could not exhaust natural evidence, and that the sphere of enquiry was rather the sphere of what was amenable to reason than the sphere of "mere" reason itself. That would be true of natural theology, for although some of the proofs of deity were supposed, as we say, to "stand to reason" (i.e. to be demonstrable from rationally self-evident premises) others were drawn in part from the observation of matter-of-fact. Supposing, however, that "reason" should be regarded as defining the level rather than as exhausting the nature of human thought, it seems fair to say that man is a rational animal precisely in so far as he is capable of weighing evidence in a reflective logical fashion. His "reason" consists of the vulgar logical powers that all men possess unless they are plainly sub-human, and is a name for the use of these powers wherever they have legitimate application. "Mixed" reason of this kind is "reason" as well as "pure" reason.

So understood, "reason" seems to be a very suitable general

Theism and Cosmology

name for man's ordinary powers of knowing in the common affairs of life. In terms of it, it seems intelligible to ask whether different *powers* are required for the common affairs of theology, or for the uncommon affairs of that or of any other science. If such higher powers are needed, "natural" theology would be foolishly and provocatively unpretentious, obstinately clutching its candle when the switch was near its hand. If, on the other hand, there is no other way of assessing evidence, or if all the other ways have too slight or too disputable an evidential value for any sober philosophy to use, "natural theology", for all its candle-light, and for all its pedestrianism, would also be both wise and prudent.

Let us attempt to make a rough inventory of a man's "ordinary" powers in a knowledgeable way—his "right use of ordinary means" to use the phrase of the *Westminster Confession*. If we do so, many items will be questionable, and none should be taken for granted. Nevertheless there may be results of some value.

Men, we say, are external observers of foreign physical bodies. In a literal sense, they may be said to have inside information regarding their own bodies, and in a sense that may not be literally spatial they are usually supposed to have introspective self-acquaintance. It is also maintained, in certain quarters, that a man's acquaintance with the minds of his fellow men involves some power additional to any of the others I have mentioned.

There are profound disagreements among philosophers regarding the powers and the interpretation of the powers that are involved in perception, in introspection and in what I may call mind-reading. The perception of foreign bodies includes sensation, memory and inference. Without inference, momentary sensation would not be evidence of permanent thinghood; indeed it may not be sufficient evidence *with* inference. Without memory in this affair we should be as stupid as Leibniz thought matter was, when he suggested that the stone in the sling feels the last little push before its release, but has too short a memory to know that it is flying off at a tangent.[1] Similar remarks apply to any developed form of introspection and also to any developed form of mind-reading, except for the notorious difficulty of deciding whether sensation plays any part, and, if so, what part in these processes. Speaking generally, however (and using traditional terms) we have some sort or sorts of acquaintance with matter-of-fact and interpret the same in perception, in introspection and in mind-reading.

These processes, as we have seen, are amenable to reason, i.e. to logic in its ordinary sense. This involves conceptual analysis—for the units of logic are propositions and propositions are always conceptualized. It also involves formal inference *secundum artem*. These formal processes can be examined in the abstract, and it is often claimed, furthermore, that certain truths "stand to reason" or are self-evident. The last of these opinions may be even more disputable than any of the others I have mentioned hitherto. Many philosophers say that what appears to be self-evident has this appearance because all experience is behind it. It does not really glow with its own light but gathers and reflects a luminous system. On the other hand, if any proposition is such that its denial presupposes itself, it is hard to deny that the proposition in question is strictly self-demonstrating, that is to say, is self-, not other-, evident.

These are among the types of ray principally to be found in the traditional "natural light" of reason, patronizingly called the "understanding" by those philosophers who believe that "reason" truly comprehended has other and grander powers. It follows that if natural theology is restricted to the "understanding", and if the understanding is reason with its wings clipped, natural theology is something perverse and self-mutilated. That may be the only circumstance that prevents its flight into the empyrean. We have therefore to ask whether "reason" has other powers.

It may fairly be claimed that the account I have given of

[1] *Philosophical Writings* (Everyman edition), p. 135.

Theism and Cosmology

the "natural light", and have called "traditional", is much too narrow to do justice to the conception. The light of reason, even when it is turned towards the plains, connotes a judicial mind, and a judicial mind has to be comprehensive as well as balanced. Vulgar logic, as well as many systems that profess to correct and to refine vulgar logic, have tended to restrict themselves to *conclusive* inferences, but there would seem to be many relevant arguments that are not conclusive. The estimation of such arguments yields one of the senses in which the word probability is used. In that sense (which includes most of the arguments that are called inductive) probability pertains to the "naturally" judicial mind. Similarly the principle of analogy has to be included, precarious as that principle may be in many ways. Induction, according to J. S. Mill, rests on the principle that there are parallel cases in Nature,[1] and I can see nothing antiquated in Mill's observation. But let me return to the requirement of comprehensiveness. Nobody pretends that a man's natural reason is *all*-comprehensive, but it would seem to me that a certain synoptic potency should be ascribed to it. The power of synopsis is not confined to poetical or to other impressionistic methods, and I do not see that it should be denied to the rational light of nature. I should say so, even if I were forced to maintain that there is imagination in natural reason. Certainly imagination is not always a truth-giving faculty, and the provenance of its gifts may remain mysterious when it does bestow truth. Nevertheless, even if scientific and philosophical imagination has to be described as a flair or as a species of vaticination, something of the sort seems to occur in most scientific or philosophical advance, even granting that the later consolidation of scientific or philosophical arguments may not require it after the first raptures of discovery are over.

The upshot of this discussion may seem to be very disheartening for any who would be natural theologians. Natural evidence, we should infer, includes the apprehension of

[1] *System of Logic*, III, chap. iii, §1.

Concerning Natural Theology

matter-of-fact, in a sense very disputable and very hard to define but indispensable in the analysis of perception, introspection and mind-reading. It includes vulgar logic, and all refinements of vulgar logic that are *only* refinements of it, and neither are nor profess to be a wholly different weapon. It includes the logic of probability and of analogy. It involves, or at least it may include, certain synoptic and imaginative powers whose boundaries are necessarily very obscure. If these statements are even approximately correct I do not see how anyone can say that the natural light is an entity whose range and scope is likely to be accurately defined. It presents its books for logical audit, but the capacities that go to the making of what is recorded in the books is very varied and is frequently obscure.

Despite these very serious objections I have to maintain, and I believe, that the sort of "natural" powers in the knowledgeable way whose general character I have attempted to indicate are genuinely knowledgeable, are more secure than any others, and may be the only relevant powers that we possess. If that be true there are very good reasons for attempting to use no others, and that, in substance, is what I believe the programme of natural theology is. Its subject is whether and what we can learn about God *in the "natural" way*, and that subject seems to me to be profoundly important. The absence of boundaries that are perfectly precise is not a fatal objection. People do see rainbows even when they cannot tell where the rainbows end. I know that the analogy is weak, but it is not utterly feeble. Passing the point, however, I should like to consider certain opposing views that may be held.

It is cheap to sneer at Hegel, but one may intelligibly dissent from the opinions of those who hold that the "understanding" should be accorded the restrained and compassionate approval that patricians give to the lower middle classes, regarding them as in a sense the backbone of the nation but also as unfitted for high office. According to this view, the "understanding" is imperfectly philosophical. Even if it suits the common affairs of life, it is very ill-adapted to the uncommon affairs of philosophy or of any philosophical theism that is worthy of the name. The criticism would be just if it could show that the "understanding" and its vulgar logic were a sort of makeshift rigged up for temporary vulgar uses and crudely simplified for the benefit of vulgar noddles. If on the other hand there is no sufficient evidence that there is any super-logic to supplant the vulgar logic, or any "philosophical" proof that proves in a different way from the unphilosophical sub-ways of common argument, that particular attack upon natural theology would not seem to succeed.

There are many ways of defending the thesis that "philosophical logic" transcends vulgar logic, and that "reason" surpasses the mere "understanding". Certain Christian and many Eastern theologians say that there is an utter simplicity in the divine nature in consequence of which the so-called attributes of deity, such as unlimited wisdom or unlimited power, fall short of the divine unity by the mere fact that they are separately discernible. These theologians commonly admit, however, that it is humanly impossible to understand the supreme simplicity of deity. Many philosophers have been bolder and have maintained that the infinite, that is, the totality of reality in its lucent integrity, is *apprehensible* in principle even by mankind, although no man can claim to have *comprehended* it in anything approaching its fullest range. Distinguished philosophers in our own country (such as Bosanquet) have held that "philosophical logic" is the spirit of totality, that there is a *nisus* towards totality in every apparent fact, that men participate in the soul of any fact merely by apprehending it, and so, in all their knowing, are wafted towards the Absolute and the Whole. F. H. Bradley's view seems to have been more simply that every judgment that does not adequately express the Whole is accompanied, by reason of its inadequacy, by an intellectual unrest that may be vague but invariably drives the mind Absolute-wards.

Concerning Natural Theology

Neither Bosanquet nor Bradley, it would seem, believed that Hegel's celebrated "dialectic" was actually a new logic which, if it were grasped, led inevitably, without any arbitrary excursions, to the goal of a self-reflecting infinitude; and McTaggart, in his latest views, seems to have thought that this goal could and should be reached by ordinary logic combined with the metaphysical principle of Determining Correspondence. Similarly when Mr. Jaspers[1] and other exponents of the new *Existenzphilosophie* complain of the ideal of Cartesian itemized step-by-step metaphysical argument, it seems to be doubtful what alternative they have to offer. What is wanted, they say, is an inclusive or encompassing philosophy instead of a philosophy that contracts itself into items of certitude; but they appear to be content to refer to "humanity" or to "history" as the source of this enveloping movement without explaining how the encompassment is effected. They seem to me to be rationalistic impressionists, less ambitious than Hegel, but not, on that account, more convincing.

I do not think it would be very surprising if the supernal overtones of reality were beyond the range of ordinary hearing, or if transfinite categories were always opposed in principle to finite categories. On the other hand, until it is shown that this should be so or must be so, I suggest that we should be very chary indeed of assuming that it is so. For myself I hope to make no such assumption, and it seems to me that any disparagement of step-by-step methods is a menace to all sound thinking. The "understanding", I allow, is not unchallengeable. The sceptics have seen to that. But if vulgar logic and ordinary reason choke you, why should you expect to be able to swallow anything else?

Another argument may be elicited from the alleged circle in the relations between "nature" and the "natural light". This supposed circle, as I remarked earlier in the present lecture, is said to be compounded of the arguments (*a*) that nature is revealed by the natural light, and (*b*) that the natural

[1] K. Jaspers, "La pensée de Descartes et la philosophie", 1937.

light is natural because it reveals nothing but nature. I shall now make some comments upon this supposed circle.

The argument used to be employed concerning miracles, and the complaint was that certain agnostics arbitrarily rejected the evidence in favour of miracles because of the presumption that miracles could not be "natural". This complaint may have some justification, although it would not upset the agnostic's conclusion that it is on the whole more probable that the so-called miracles had been mis-reported than that they had actually occurred. It would certainly be a vicious circle to argue to-day that the evidence in favour of super-normal spiritualistic phenomena should be disregarded because it deals with what it professes to deal, viz. something supernormal. Clearly, however, there is something wrong with the proof of circularity upon these lines, for most of the evidence in favour of miracles, prodigies and the like is ordinary "natural" evidence. No extraordinary power of perception was needed to discern by taste that water had changed into wine. There is no contradiction in saying that man's ordinary powers may reveal what is not ordinary.

A much more formidable argument, with which we shall frequently be occupied in the sequel, is that the principles of natural knowledge apply *within* nature and not *to* nature, with the consequence that natural theology, even if it assumed that God and Nature were identical, could not reach any natural knowledge of its subject.

Of this argument I shall here say, briefly, that even if natural theology deals expressly with an object wholly without parallel, it need not, on that account, require a logic, a percipience or other knowledgeable faculty that is also wholly without parallel. I would add that if some of the principles of natural knowledge apply *within* and not *to* nature it is rash to assume that all of them are hobbled in this way, and that the view that there could be no intimations of deity within nature (so that all such enquiries are forbidden) may well be unreasonable.

Concerning Natural Theology

It is sometimes maintained that "religious experience", to return to that topic, attests the truth of divinity, but is of a different order of knowledge from what we call the knowledge of nature. In the alternative, it may be said that nothing is more natural to a man than worship, and therefore that the natural clarities of devotion should not be excluded from the light of nature on the mere ground that they are irrelevant to the baking of bread or to the computation of parsecs. Men are made for Jerusalem, not simply for Babylon. Here our essential question is whether religious experience includes any knowledgeable powers that are absent from other experience. Men do have religious experience. But do they have a theopathic sense additional to their visual, auditory and other senses? Is any other peculiar mental faculty present in religious experience and in it alone? Is divine wisdom, beauty or goodness discerned by a different organ from terrestrial wisdom, beauty or goodness? Is the sunlight of grace different in kind as well as in power from the candlelight of natural reason?

When this question is candidly put we are entitled (I think) to reply that whatever the true answer may be, it is legitimate (to say no more) to proceed upon the hypothesis that an affirmative answer is too doubtful to be a reasonable foundation for philosophical argument. Neither the sentiment of reverence nor what is called communion with God is unambiguous evidence of the existence of a peculiar theopathic faculty, and if these instances do not prove the desired conclusion I do not know what else in "religious experience" could conceivably do so.

Most certainly the sentiment of reverence exists and is profoundly moving, but it is not non-natural and it need not apply to, or imply the existence of, a non-natural object. It occurs (does it not?) in the presence of anything that we deem sublime. It may be shown towards greatness in mankind, and sometimes towards a very earthy sort of greatness. Mountains and vast plains, the stars and the sea may induce it. Wherever there seems to be majesty there is sure to be awe. **Natural**

Theism and Cosmology

piety of this kind may contain vaticinations of deity but it is neither exclusively nor obviously theopathic.

Of "communion with God" I can speak from hearsay only. If a man believes that he walks with God, that God responds to his heart and mind, in brief that there is genuine although infinitely unequal fellowship between deity and himself, he is claiming, in principle, a personal acquaintance with the divine being not wholly dissimilar from his acquaintance with other men. We have therefore no need to assume more than the heightening of a natural power; and how can it be denied that the greater claim, however passionate its sincerity, may nevertheless be mistaken? Again, how could anyone know that if God the Father were so revealed, he was also the creator of heaven and earth?

Accordingly I submit that the "right use of ordinary means" of knowing has an intelligible if not a precisely demarcable sense, that it is doubtful whether we have any other means of knowing, and that it need not be true that the knowledgeable powers we apply within nature could not apply to nature or to nature's God.

II

Concerning Theism

IN the present lecture I mean to discuss the meaning of the term "theism" in the hope of clarifying our ideas before going on. An appreciable advance would be made if the dangers of an over-flexible and of too narrow an interpretation of the conception were adequately understood. Such a discussion, I think, is much more than verbal although its nominal subject is the way in which a word is used.

In literary usage "theism" is a younger term than "theology", and it may denote a narrower because an exclusively positive field. If by "theology" one meant a reasoned discourse concerning the things that are called divine, the negative side, i.e. a reasoned atheism, would be a kind of theology. And atheism is certainly opposed to theism.

That, however, is a small matter. It is more significant to examine whether the term theism is itself ambiguous, and, if so, in what principal ways. We may conveniently begin by examining the common usage of the term in the English language.

In its general sense, which, according to the *Oxford Dictionary*, is good seventeenth-century English, theism is simply the opposite of atheism. This very general sense seems to me to be also the most convenient, and I intend to make a beginning with it. In the widest way, the meaning is that anyone who asserts that there is a God or that there are gods, whatever positive meaning he may attach to the conception of Godhead, is a theist. If his conceptions of deity were feeble or vague, his theism would be feeble or vague, but still it would be a species of theism.

Theism and Cosmology

The *Oxford Dictionary*, however, recognizes two narrower senses of the term "theism". In the first of them "theism" means monotheism and is said to be opposed both to polytheism and to pantheism. In the second it is opposed to a mere deism.

Of these narrower senses, the first seems to me to be reprehensible in a philosophical way. Monotheism is a term that does not need improving. Polytheism professes to be a form of theism, and should not be extruded or contemned by mere definition. Again, although pantheism, in some of its senses, may be interpreted as the most extreme form of polytheism that is conceivable, it notoriously has other senses in which the unity as well as the divinity of the All is most emphatically affirmed.

The second contrast is of greater moment. It is often useful to have a term that explicitly avoids the nakedness of mere deism, but it seems to me that justice and expediency are opposed to the view that deism is not a form of theism. There are, in the first place, historical difficulties of a linguistic kind. The *Oxford Dictionary* assures us that "theism" and "deism" were convertible terms in late seventeenth-century English. That might not matter, but even to-day, according to the same authority, there is a tendency to interpret deism in a sense that could not be ignored by a lecturer on natural theology. A deist, in this dictionary, is defined as "one who acknowledges the existence of a God upon the testimony of reason, but rejects revealed religion", and this dictionary, very cautiously, defines theism partly in the negative, i.e. as "belief in one God as creator and supreme ruler of the universe *without denial of revelation*".

In short, as a matter of usage, deism is frequently regarded as a form of theism and not as a denial of theism. I propose to follow this usage without, of course, objecting to the procedure of those who prefer to distinguish the terms by definition, and who stick to their definitions. The usual philosophical distinction of this order is that a theist asserts God's providence,

Concerning Theism

goodness, and, as nearly as may be, the personal quality of his spiritual being, whilst a deist holds himself aloof from all such assertions unless an icy intellectualism be deemed their partial equivalent. In this sense Mr. Taylor[1] argues that Aristotle was the father of deism while Plato was not only the ancestor of philosophical theism but also (in the tenth book of his *Laws*) was himself a theist. There is here no question of accepting, rejecting, or neglecting revelation, and no internal contradiction or other impropriety in a theism that confines itself to philosophical arguments.

A theist, then, in the language I propose to use, is one who holds that there is a God, or that there are gods, in any of the forms in which such beliefs may be held. An atheist denies the existence of a God or of gods in all possible senses. The two terms are opposed, and are not bare logical contradictories, because there is another alternative, namely agnosticism. An agnostic asserts that we do not know, or, in the extreme form of his theory, that we never could know whether there is a God (or gods) in any sense whatsoever.

It is plain, however, that "theism" in this general sense might be puerile, vague, and sentimental, as well as serious and philosophical. The point may be illustrated from the speech of Socrates at his trial as recorded by Plato. Socrates asserted that he could not be an utter atheist, if, as the accusation stated, a part of his alleged impiety was an attempt to introduce new and strange divinities into Athenian worship. That argument was good enough to embarrass Meletus, his clumsy-witted accuser; but it is plain that Socrates neither intended nor desired it to fool the jury. Socrates had little interest in a civil offence of that kind, or in a ceremonial punctilio. When, as his speech approached its crisis, he argued with all his soul that his mission, his duty and his inhibitions were divinely inspired, he spoke of another sort of theism, the theism of deep philosophy.

The same holds for us. We are not here concerned with

[1] Art. "Theism".—*E.R.E.*

Theism and Cosmology

every type of God or of godling that men have honoured,[1] high gods and low gods, nature gods, *Sondergötter*, tribal gods, all that is daemonic or numinous, divinized heroes and ancestors and abstractions, the legends of the poets and the theogonists, or even what Coleridge called the "fair humanities of old religion". Therefore we should attempt to indicate, if we can, some of the major requisites of a theism that can properly be called philosophical. At the same time we have to beware of fixing the limits of philosophical theism in such a way as to exclude views that are relevant and that also are capable of reasoned philosophical defence.

Consider, first, the lower limits. Philosophical theism cannot be altogether naïve, but a naïve theism may naïvely affirm a view which, after much philosophical pondering, may seem to be a crystal verity that has to be accepted *sub persona infantis*. More generally, nature-worship, the cult of heroes, supernal guidance in history, the conservation of values, and the like may be receptive of philosophical deiformity despite the infantile, repellent or ingenuous ways in which some such beliefs have sometimes been accepted in pre-philosophical theologies. It is the apex of such a pyramid and not simply its base that concerns us. The better our philosophy, the better it should be able to descry the promise and potency of groping and primitive ideas.

Consider, again, the upper limits. Dr. Tennant has recently complained,[2] rather unkindly, of what he calls "the more pretentious *a priori* theology". He was referring, in the main, to the doctrine of the divine attributes as developed in Christian theology, and I should like to follow him in deprecating "premature assumptions" in that matter. It is part of the business of natural theology to examine these attributes, and to gain immensely (whatever its conclusion about them may

---

[1] "Worshipping snakes or trees, worshipping devils rather than nothing; crying for life beyond life, for ecstasy not of the flesh. Waste and void. Waste and void. And darkness on the face of the deep."—T. S. Eliot.

[2] *Philosophical Theology*, II, 122.

Concerning Theism

be) from the care and the genius that has been spent upon elaborating them. If, however, we laid down in advance that no theistic theory was to be taken seriously unless it dealt with a God defined by these attributes, or by these attributes as united in an utter simplicity, we should be in danger of an enormous loss. There may be many forms of philosophical theism that escape such treatment or are uneasy in its presence.

In short, we should understand our subject in a way that leaves it free for all legitimate self-development. On the other hand, some attempt to indicate if not actually to fix the lower limits of a philosophical theism should not be avoided. I shall suggest and discuss three such limits. It seems to me (*a*) that no theism is of a philosophical order unless it holds that divine power operates effectively upon a cosmic scale. Philosophical theism must be at least diacosmic. I hold (*b*) that philosophical theism implies an effective unity in its conception of divinity, whether it be monotheistic or polytheistic. Again, although more doubtfully, I should like to suggest (*c*) that any so-called theism that does not profess to be ultimate is not, in the strict sense, theism at all.

Let us consider these questions *seriatim*.

(*a*) Crude theism makes a gesture towards cosmology when ıdiscerns its deity in the thunderer, believing him to rule the tempest and direct the storm, or when it finds the source of all fertility in him. Its horizon, when it discerns these things, is earth- or sky-bound. Nevertheless, the powers supposed to be thus discerned have the marks of what may be called an earthy infinitude, and strain after a primitive majesty. It would be a mistake, it is true, to call the powers so conceived non- or super-natural, in the modern sense. They were not super-scientific. Human beings, at the time when they framed these cruder theistic ideas, did not distinguish as we do between what is and what is not scientifically verifiable. In that sense they were unscientific, but they need not have been alogical in any fundamental way. They reckoned with powers that *we* should describe as non-natural because they are remote from

Theism and Cosmology

all *our* telescopes and from all *our* test-tubes, and they may have interpreted such powers very largely in terms of what *we* now call animatism, animism or magic. They intermingled conceptions of that type with what we should call science and common sense. They included augury in the art of war as well as shields and horsemanship and poisoned arrows. If, however, we consider generally what they believed, as opposed to what they should have believed had they assimilated a modern scientific education, we have no good reason for denying that they thought of divine power as something indefinitely stronger than a man's hands or than the waves of the sea.

It is relatively unimportant, no doubt, to conjecture what a crude theism may have crudely intended. The essential thing to note is that in any theism that can be taken in philosophical earnest the emphasis upon cosmology is quite fundamental. "The theme of cosmology", Mr. Whitehead has recently said, "is the basis of all religions."[1] His statement is not quite accurate. One thinks, for example, of what Sakyamuni the Buddha said about the finitude or infinity of the cosmos. "These questions are not calculated to profit, they are not concerned with the Dharma, they do not redound to the elements of right conduct, nor to detachment, nor to purification from lusts, nor to quietue, nor to tranquillization of heart, nor to real knowledge, nor to insight of the higher stages of the Path, nor to Nirvana. Therefore is it that I express no opinion upon them."[2] On the other hand, although *religion* may abjure cosmology, I would suggest that *theism* dare not do so. To a theist cosmology is inevitably a major theme. I do not say that it should be altogether dominant in theism. For God may be hyper-cosmic as well as dia-cosmic and the former aspect may be prepotent over the latter. But God cannot be less than cosmic in his scope and power.

We need not (and I think we should not) say with Hobbes that "power irresistible justifies all actions, really and properly, in whomsoever it be found: less power does not: and because

[1] *Process and Reality*, p. 494. [2] Suzuki, *Essays in Zen Buddhism*, p. 125.

Concerning Theism

such power is in God only, he must needs be just in all his actions, and we that, not comprehending his counsels, call him to the bar, commit injustice in it".[1] On the other hand, if God (or the gods) were well-meaning, or just, or tender, or compassionate—*but powerless*, theology would be a feeble affair. Who would impute divinity to an *ineffective* bounty, or to a tender mercy too gentle to redeem? Isis, it was held, could "save the universe". If Calvary had been the end of the Incarnation it would have been the end of a sad human story. Christ must come in triumph and with power in order to assert his divinity.

I am not suggesting that there could not be a *religion* of humanity. Much in humanity may be worthy of reverence, for much in it is fine and pure and saintly. It may even be perfectible in its kind as the Christian doctrine of the Incarnation proclaims in a figure. The nature of man may be worthy of reverence, but mankind cannot be worshipped *as divine* for the simple and sufficient reason that neither individual men and women, nor humanity collectively, are effective agents on a cosmic scale. That is apparent from one of Comte's own statements, in the preface to his *Catechism of Positivism*. "In the name of the Past and of the Future", he there said, "the servants of humanity—both its philosophical and practical servants—come forward to claim as their due, the general direction of *this* world. Their object is to constitute at length a real Providence in all departments, moral and intellectual and material." That may be a hypothesis to live by and to die for, even if men and Man be in the end ephemeral. It may be admitted to describe a pragmatist's Providence, if by Providence one means nothing more than the work of redemption in mankind during so long a season as human beings may survive. What I am asserting is that it is not *theism*. It has nothing to say if man's destiny is to perish or, still more tragically, to deteriorate slowly but never quite unto death. Theism, because it is at least dia-cosmic, makes quite other

[1] *English Works* (Molesworth), V, 116.

Theism and Cosmology

assertions about God whatever it may have to say about the destiny of mankind.

The same is true, I suggest, of any philosophy that, in place of a world-theory, or of something greater than a world-theory that yet includes a world-theory, is content to affirm the reality of spiritual oases in a natural desert which itself is conceived after the fashion of positivistic science. Such philosophies are atheistic although they need not be irreligious. According to such views spiritual communities, pagan or churchly, cultural or racial, artistic or barbarian, may indeed exist and have a certain vital integrity. Each, according to Spengler,[1] is a "second cosmos" contrasted with the "first cosmos" of Nature. So of the spirit of any particular man; and it is correct to say that a nation, a calling, an institution and (in general) any collective body may be nobler, stronger, and greater than its component members. No such collective human entity, however, is strong enough to be a God. Even if its spiritual quality were divine, its spiritual reality would not be the Godhead. If its relative strength and integrity were of the same order as the brief integrity of a human spirit, lasting as a flower lasts when sunlight and moisture are propitious, it could not be truly divine. A spiritual zone in an unspiritual waste, a cultural *Gau* in an astronomical desert, is only accidentally strong or lasting unless it has the backing of a greater, more God-like power. While it lasts it may have the fineness and the beauty that divine things retain, but its weakness is a denial of its divinity. It cannot be content with extra-territorial rights in the Cosmos.

Power on a cosmic scale, therefore, appears to me to be implied in any seriously philosophical theism. By the cosmos, for the purposes of this statement, I mean primarily the physical universe; for cosmology, as ordinarily understood, is concerned with that universe. I am not maintaining, now, that reality may not be richer than the physical universe, that

[1] In *The Decline of the West*. (London: George Allen & Unwin Ltd.)

Concerning Theism

human and even animal nature is not more than physical, that Nature may not be hyper-physical, or, for that matter, super-spiritual. I am not asking, now, what the physical universe is. I know that philosophy has, and will continue to have, a great deal to say upon that perplexing subject. In short, I do not want to prejudge any large philosophical questions. What I am maintaining is simply that theism must at the least be cosmological. If it professes to be less than cosmological, it falls short of genuine theism.

Some further explanations should be appended. I have said that if theism be true, God (or the gods) must operate effectively upon a cosmic scale. This statement, when examined more closely, should be understood to mean that deity, according to a theistic philosophy, plays a *large* part in cosmology. Indeed, it may be doubted whether there is theism, properly speaking, unless the prepotency of divine action in cosmology is asserted. In any case a mild but ubiquitous divine influence in cosmology would not be enough to satisfy a theist. On the other hand, I am not maintaining that unless God's cosmic power is limitless, there can be no theism. A limited deity or a limited company of limited deities is a possible theological hypothesis. It would be absurd to exclude such views by a rash preliminary definition, to hold that we must either be atheists or say with the Bhagavadgita:[1] "Thou art the Father of this world, of all that moves and does not move; thou art to be adored". But anyone who holds that cosmology is irrelevant to theism, neglects what is quintessential to every theistic philosophy.

(*b*) In the second place I would submit that theism is either sub- or non-philosophical unless it ascribes a genuine unity to its deity.[2] I shall discuss this point, briefly, with further reference to the distinction between monotheism and polytheism and to some of the questions that arise in pantheistic theory.

---

[1] XI, 20.

[2] Since evil is often lacking in unity the subject of polydaemonism *versus* monodaemonism does not run parallel to polytheism *versus* monotheism.

Theism and Cosmology

It is for anthropologists to decide whether a crude or a naïve monotheism does not always (or very frequently) precede the polytheism that may be slightly or greatly more cultured. That is not a question that concerns philosophy, any more than the puzzle why, if monotheism is more readily defensible than polytheism, it should have succumbed to polytheistic criticism in so many cultures. In general, however, a very wide dissemination of divine functions, or a scattered distribution of hallowed powers in sacred places, suggests a diluted and sprawling theism too weak for philosophical health. There would be altogether too many little gods, pigmy godlings, few of whom (or of which) deserved the name of deity; and the view that there are just a few genuine deities (say, a quota of half a dozen or a pantheon of nine as at Heliopolis in Egypt, or a triad like Brahman, Vishnu, Siva) seems fantastically arbitrary. A limited deity, it is true, must be limited by something unless he limits himself. But he need not be limited by other *gods*, and if he be limited, as on the Manichaean theory, by an organized opposition, by a united, independent and quasi-personal power of darkness or of evil, such a being need not really be a god.

Nevertheless, it should not be assumed prematurely, and perhaps it should never be claimed, that among the higher and more philosophical varieties of theism, monotheism is invariably at the top. Philosophical debates concerning the One and the Many have their application to philosophical theology, and affect the interpretation of monotheism as well as of philosophical monism. Suppose a theist's conclusion to be that a unified spiritual power has dominion over the universe. If so, our theist might still be entitled to believe that this unified spiritual power was a spiritual society, a collective super-mundane force with all the high degree of unity that a (superhuman) society can muster. Some would say, it is true, that such a theory is not theism but an alternative to it. If they say so, I should regard their assertion as thoroughly arbitrary.

Concerning Theism

What Kant called the "kingdom of ends", and described as a *corpus mysticum*, might also have been called a noümenal republic, or, again, might have been described as a society which, like ideal communism, had transcended the need for any sort of hierarchical or political government. Indeed Kant might have said so had he not so plainly been thinking of Leibniz who described the kingdom of heaven as "that perfect commonwealth of minds which deserves to be called the City of God";[1] and Leibniz retained monarchical language principally because he held that God had ordained the harmony of the monads, like a clock-maker who has synchronized all his clocks forever, yet without himself being a clock among the others. Again McTaggart, a great interpreter of Hegel, held that Hegel's ultimate theory was reached in the assertion that the Idea had attained perfection when it "took up its abode in the Church", and that "real spirit" or "God as Present" was "God dwelling in his Church".[2] If so, it would be a verbal question whether theism should be defined singularistically as the doctrine that God is *one* divine entity. McTaggart himself argued that there was no God in the singularist's sense, that the singularist's conception was of little religious value, and that the singularist's reasons, if sometimes plausible, were plainly vincible.[3] There, I think, he was fighting a mimic battle.

Let me turn to pantheism.[4] This theory has two principal forms, not necessarily inconsistent, but readily distinguishable. The first may be called distributive pantheism. It was stated, long ago, in a fragment attributed to Hippocrates. "Nothing is more divine or more human than anything else, but all things are alike and all are divine." It may also, however, be totalitarian, and that may have been what Xenophanes, another pioneer of the Greek Enlightenment, meant when he said that the All, or the "world", was God.

[1] *Philosophical Writings* (Everyman), p. 236.
[2] *Studies in Hegelian Cosmology*, p. 208.
[3] *The Nature of Existence*, II, 183.
[4] I prefer the general term and so have not spoken of panentheism, theopanism, etc.

Theism and Cosmology

The primary intention of both these forms of pantheism may have been to de-theologize rather than to assert any form of theism, and this view, as regards the Greek Enlightenment, may be supported by the recollection that the word θεός, in the Ionian dictionary of natural science, frequently meant what was elemental in physics and nothing besides. The medical school of Cos, in particular, with its "Hippocratean" literature, was concerned in the main to extrude cosmology from medicine, and to follow the way of experiment. A famous early statement referred to the case of epilepsy. "Men fancied it and its cause supernatural from inexperience and surprise at its difference from other disorders. . . . My belief is that those who first called this disease 'holy' were men such as we see to-day, wizards and purifiers and mountebanks and quacks. . . . They made the supernatural a cloak for their own helplessness and inability to produce a remedy, and reckoned this disorder holy to prevent the detection of their own want of science."[1]

Whatever these early authors may have intended, however, it is just as legitimate, in a speculative way, to attempt to sanctify all that is secular as to try to secularize all that is sacred. That, as I should suppose, is what pantheism endeavours to do whether it be distributive, or totalitarian or both; and pantheism, so interpreted, seems to me to be obviously a form of theism. In its distributive form, such a pantheism would be pluralistic. In its totalitarian form it would be monistic. Neither form need be what is currently called monotheism, but neither would be a serious contribution to philosophical theism unless it asserted the unity of deity in a high degree.

The identification of God with Nature has not been uncommon. "Quid aliud est natura quam deus?" Seneca asked.[2] "Jupiter is whatever you see, and whatever movement you notice", said Cato in *Lucan* (IX, 580). In the fourth century

---
[1] Trs. Taylor, *A Commentary on Plato's Timaeus*, pp. 602 ff.
[2] *De Benef.*, IV, 7.

Concerning Theism

Aetius boldly said that "we know God as we know what we see with our eyes or handle with our hands—as we know a stone or a log or any other material thing".[1] The last of these statements may reasonably be regarded as even more objectionable than the doctrine of a wholly "unnatural" God. But let us skip a millennium. Spinoza's famous phrase *Deus sive Natura* may indeed be supposed to naturalize deity, but *Natura sive Deus* would divinize nature, and the latter, surely, was at least equally in Spinoza's mind.

That was Hegel's comment when he said that Spinoza's crime was acosmism rather than atheism.[2] The world, for Spinoza, was caught up into deity. In its temporal march it had only imaginative (not necessary) being, and although Hegel himself was hostile to pantheism in the sense in which he professed to define "pantheism", his own panpsychism was pantheistic in its own way and must be allowed to be a form of philosophical theism. Obviously, Hegel was a totalitarian pantheist, for he held that "without the world, God is not God", thus implicitly identifying God with the Whole. Certainly the associative penumbra of the word "God" in the vernacular is transcendent rather than immanent. Therefore if it seems to be desirable (or even to be but candid) to refrain from perturbing this associative penumbra when we speak of God, it may be expedient to decline to describe pantheism as a form of theism. I think, however, that this would be a mistaken policy as well as a sin against the logic of the situation. Totalitarian pantheism, like the Absolutism that transcends vulgar theism, might be regarded as the philosophical truth that theism misses but comes nearer to asserting than any other unphilosophical belief. It seems to me to be more accurate to regard it as a possible form of theism, that is, as theistic just because it is pantheistic. In any case it is desirable to take this course when, as at present, we are trying to allow sufficient although not exaggerated elasticity to the play of philosophical argument.

[1] Quoted by Kirk, *The Vision of God*, p. 306.
[2] *The Logic of Hegel*, Trs. Wallace, p. 106.

Theism and Cosmology

(c) It seems to me that every theistic philosophy must, in intention, be an ultimate philosophy. No doubt, theism may be put forward, tentatively and provisionally, as a hypothesis. Even in that case, however, it should always be put forward as an ultimate hypothesis. If the gods are so conceived as to require an ancestry, the conception falls short of philosophical theism. Indeed, as I understand the question, it is just because philosophical theism, from the nature of its claims, is an attempt at *ultimate* explanation, that it is ineluctably metaphysical or philosophical. It is one of the ways of attempting to deal finally with first things and with last things, and philosophy, for that reason alone, is bound to take account of it, not simply as a contribution to philosophy, but also as itself a philosophy.

These statements, it is true, need a word of comment. It would not be true to say that every form of theism, in the past, has been, *inter alia*, a piece of would-be ultimate metaphysics. The idea of a nation-God, for instance, with whom a tribe might enter into some sort of compact, is *not* a piece of ultimate metaphysics, and yet may have represented a relatively advanced type of theism. Turning to philosophy, again, we may readily allow that many theistic systems might be appropriately described by philosophers, not as something ultimate, but, at the most, as *phenomena optime fundata*. What is really ultimate, some philosophers hold, is the Absolute, and God (they say) is not the Absolute. He is only the best-grounded of all appearances, the least misleading way of approaching the ultimate in figurative language which, because it is but figurative, is *eo ipso* not ultimate. Indeed it is not impossible that unphilosophical theists might have no quarrel with such a philosophy, holding that it gave them all the Godhead that they wanted. I am talking, now, about philosophical theism, the sort of theism that is in earnest with its philosophy, not about the sort of theism that is pre-philosophical or about the sort of theism that is content to let philosophers wrangle outside its own closed doors. What I am asserting is that

Concerning Theism

philosophical theism is bound to regard deity as an ultimate explanation. If it did anything else, it would really be denying the truth of theistic philosophy and would be saying instead that theism is an imaginative gloss, perhaps very profound, upon something that, really and strictly understood, is not theism at all.

Again, from the cosmological standpoint, there have been many cosmological theistic theories that were not ultimate. Many Neo-Platonists distinguished the Demiurge of Plato's *Timaeus*, the "father and maker" of the moving and visible universe, from the ultimate rationality of the One. If the universe were the changing image of an eternal rational model, and if the "father and maker" of the heavenly gods, i.e. of sun, planets and earth, were *only* a creator and *yet* the highest god, the ultimate truth would go beyond theism. To choose another illustration, certain of the Gnostics held that nature was the handiwork of an inferior type of deity, perhaps of the devil. It could not be the product of the great "unnatural" God, since it was opposed in all its being to the standards of true divinity.

Thus various forms of theism may be and have been understood to be *not* metaphysically ultimate. Nevertheless, as I have said, it seems to be part of the meaning of philosophical theism that it is final. It is a doctrine that tells us where we ought to stop, not merely a doctrine of some convenient resting-place, and I mean to treat it so in the sequel.

I think, however, that I may help to avoid misapprehension if I illustrate this conception of the office of philosophical theism by discussing certain cases in which there may be a good deal of doubt regarding what is and what is not to be accounted "ultimate" in a philosophical way. For this purpose I shall make some comments, firstly, upon the theory of a limited deity, secondly upon the theory of a "God in the making", and thirdly upon the theory of an "emergent God" in so far as this third theory may be distinguished from the second.

Firstly, then, the theory of a limited deity.

Theism and Cosmology

If God, although supreme, be not the whole of reality, he is, in one sense, "limited", and again may be called "finite" in one of the senses of that term. Again, if he were the whole of reality he might, perhaps, be self-limiting, although he could not be other-limited, that is to say limited by anything *else*.

Monistic philosophers maintain that the whole in its wholeness is alone ultimate. Hence they infer that theism cannot be an "ultimate" theory in the metaphysical sense if it holds that God, even if supreme, is not the whole. I should like to repeat, therefore, that nothing I have said is intended to prejudge the issue between the metaphysics of monism and the metaphysics of pluralism. If philosophical theism be put forward as an ultimate truth, and if it is not a form of pantheism, a pluralistic metaphysics is implied. I am assuming for present purposes that such a metaphysics may be defensible. If so the theory of a limited God may well be an ultimate theory. It may not be the best such theory, but it would be a type of theism, and it would be susceptible of many different forms. If God were wholly good, and had to meet the united forces of evil (these together comprising the whole of reality) that would be a type of ultimate theory, although it might not be very convincing. If, besides God, the universe contained scattered and parasitic evil forces, or recalcitrant facts that, without being evil, hampered divine goodness, that also might ultimately be true. The same would happen if God had to work upon an uncreated space, "that shady nothing out of which the world was made". If we denied *in limine* that such theories should even be entertained we should be doing violence to the entire discussion.

That is what I would say, at this preliminary stage of our enquiry, about the theory of a limited deity, where the limitation is held to be due to something *else*. As regards self-limitation we shall have abundant opportunities for seeing that the mere removal of all apparent "limitations" from the divine nature raises at least as many difficulties as it professes to solve.

Secondly, the idea of a God in the making.

Concerning Theism

According to this theory, God evolves as well as his universe. We may think of him as Omega rather than as Alpha, but we should never think of him as a *completed* Omega. According to this theory, we ought to contemplate a *growing* deity, one who resembles all the living things that are "dans l'enfantement d'eux-mêmes". Such views, as was only to be expected, have become more frequent in post-Darwinian times. We are becoming, it may be, more faithful to the spirit of history. There were also, however, pre-Darwinian believers in a growing God, especially among metaphysical romanticists. Thus Schelling said very early in the nineteenth century: "Has creation a final goal? And if so, why was it not reached at once? Why was the consummation not reached from the beginning? To these questions there is but one answer: Because God is Life and not merely Being. . . . The philosophy of nature is in its most comprehensive sense cosmogony, or, as Moses called it, *Genesis*".[1] Again, he said in 1812, "I posit God as the first and the last, as the Alpha and Omega; but as Alpha he is not what he is as Omega, and in so far as he is only the one—God 'in an eminent sense'—he cannot be the other God, in the same sense, or, in strictness, be called God. For, in that case, let it be expressly said, the unevolved God, *Deus implicitus*, would already be what, as Omega, the *deus explicitus* is."[2]

It may reasonably be debated whether a bio-theism of this kind is theism at all, and whether, if theism can become a sort of meta-biology, it must not, in so doing, transcend biology and evolution altogether. Alexander himself expressed his doubts. "Deity", he said, "is a nisus[3] and not an accomplishment. This, as we shall note, is what prevents the conception from being wholly theistical."[4] On the next page he said, "God as an actual existent is always becoming deity but never

---

[1] Quoted by Lovejoy, *The Great Chain of Being*, pp. 318 and 320.
[2] Ibid., p. 323.
[3] A "nisus or ἔφεσις", erudite people say, preferring the decent obscurity of a learned language. [4] *Space, Time and Deity*, II, p. 364.

Theism and Cosmology

attains it. He is the ideal God in embryo. The ideal when fulfilled ceases to be God, and yet it gives shape and character to our conception of the actual God, and always tends to usurp its place in our fancy."[1]

Many philosophers would say that eternity and vicissitude are opposed, and that (if we have to choose between these opposed orders) God expresses the aspect of eternity and not the aspect of vicissitude. Their inference is that the conception of a growing God is atheism rather than theism, if it be intended to deny God's immutability. That is an argument about the metaphysics of temporal process as much as an argument about theism, and the topic is much too big for to-day's lecture. What I should like to say here and now, quite simply, is that *if* it is possible to maintain, on metaphysical grounds, that temporal process is ultimate and that the sense, if there be one, in which we may properly speak of "eternity" does not affect the truth of this contention, *then* there need be nothing "imperfectly theistical" in the conception of a "God in the making". The argument would then be that everything that exists is in the making, and so that God, if he exists, is in the making too. If that were true, many theistic philosophies would be denied; but I think it would be arbitrary to assert that no form of theism could be retained.

I say this notwithstanding the fact that some philosophies have attempted to search for ultimacy beyond existence itself. Here is Mr. Nicholson's[2] translation from a mystic of Islam:

> I died as mineral and became a plant,
> I died as plant and became animal,
> I died as animal, and I was a man.
> Why should I fear? When was I less by dying?
> Yet once more I shall die as man, to soar
> With angels blest: but even from angelhood
> I must pass on; all except God doth perish.

[1] *Space, Time and Deity*, II, p. 365. [2] *The Mystics of Islam*, p. 168.

Concerning Theism

> When I have sacrificed my angel soul
> I shall become what no mind e'er conceived.
> O! let me not exist! For non-existence
> Proclaims in organ-tones "To Him we shall return".

I may add another observation. If it were said that a developing God would be only time's creature and therefore not a God, and that he would be subject to time, I think it might fairly be replied that all this talk about creatureliness and about subjection may be much more verbal than formidable. If everything that exists is temporal, temporality is not a limitation of existence, but an essential property of all existence. To deny it, while accepting the premiss, would be to say that something (for example, God) existed without existing.

Thirdly, the theory of deity as an "emergent" quality.

The theory of "emergence" has been illustrated from Browning's *Abt Vogler*:

> I know not if, save in this, such gift be allowed to man,
> That out of three sounds he frame, not a fourth sound, but a star,

although without Browning's interpretation of the marvel, namely that "the emulous heaven yearned down". Novel properties are said to "emerge" when certain constellations or collocations occur, and are called "novel" when they cannot be deduced from the collocated constituents. They are *products* not mere *educts* as Kant said when he contrasted epigenesis, as he called it, with preformation. Hence it is argued that life emerges from the inorganic, not because of a new vital substance, but because a certain collocation of matter displays the "emergent" property of being alive. Similarly mind may "emerge" from life, i.e. a *minding* living body is evolved.

Granting that a whole may have distinctive properties that its constituents do not severally possess, it would seem to me that this doctrine of emergence tends to be employed in a way that is much too ambitious for its apparently innocent structure. What is supposed to be the historical emergence of

Theism and Cosmology

life from the inorganic and of mind from unminding vitality is treated as a paradigm of all emergent process with the result that emergence is taken to be the technical name for an emergent *progression* in time. For this the general conception of "emergence" seems to me to supply no warrant. There is no general reason why "emergence" of this kind should be the marrow of all history, and none that forbids retrogression. It is not at all necessary to "think of Time as a friend".

Suppose, however, that there were a general law of *progressive* "emergence" in reality, and so that the world could be said to be straining after something higher than mind. The question I now want to ask is whether, on this assumption, a theist could legitimately draw the inference that Mr. Alexander draws in the following passage, "Within the all-embracing stuff of Space-Time, the universe exhibits an emergence in Time of successive levels of finite existences, each with its characteristic empirical quality. The highest of these empirical qualities known to us is mind or consciousness. Deity is the next higher empirical quality to the highest we know."[1]

There is at least one respect in which some such conception as this may be held to be superior to vulgar theism. According to many theologians, vulgar theism greatly exaggerates the extent to which God's spirit should be regarded as a supreme and magnificent *mind*. For this reason M. Maritain has recently asserted that when natural theology elected to produce its theodicies it "religiously prepared the way for atheism".[2] Hence, we may be told, it is all gain if we learn that deity differs in kind from what we call mentality, just as clearly as mind differs from mere life; and although we cannot hope to discern what is above our own level of comprehending, we may still be able to discern the principle although not the character of the progression.

My difficulty concerns the statement that deity is the *next* quality that is higher than mind. If that statement were true

---

[1] *Space, Time and Deity*, II, 345.
[2] *The Degrees of Knowledge* (English translation), p. 278.

Concerning Theism

there would be no reason why deity should not be succeeded (either temporally or logically) by an emergent quality still higher than deity, and so on without end. If *we* should worship "God", "God", in his turn, should worship something better than "God", and that super-deity should also be a worshipper of his immediate superior. It does not seem to me that such a conception is genuinely theistic. It may be true that vulgar theism does not get beyond the next stage in such a progression if, indeed, it gets so far; but why should a philosophical theism stop at the next stage? Dare it stop short of the top? And if, as on this theory, there is no top, is there then such a thing as theism?

I have now attempted to give a general account of the meaning, and of the upper and lower limits of theism in its most general sense. In doing so I have taken the risk of being too hospitable in my interpretation and also of being stingily inhospitable. I have tried, however, to be as hospitable as I could, not because philosophy should always keep an open house, but because heresies are better to hear than to hunt.

More in detail, I intend to treat theism in these lectures as a theory that professes to be metaphysically ultimate, and is also at the least cosmological. For the sake of emphasis I shall allow myself the argumentative extravagance of repeating these contentions in the sense that I should like to put upon them—I hope, not too dogmatically.

What is ultimate, in the sense that is relevant here, is what is not derivative and is not provisional. According to theism God is not logically derivative. The proof of him is the proof of a *prius*. He is not genetically derivative, for he is regarded as *first* existence, *natura naturans*, not *natura naturata*. When he is regarded teleologically, he is regarded by theists as *finis ultimus*. The theistic hypothesis, it is true, may, like any other hypothesis, be put forward provisionally or tentatively, and so may a theistic *demonstrandum*, but neither hypothesis nor *demonstrandum* can intelligibly be provisional in the sense that if they were established they would only show the existence

Theism and Cosmology

of something provisional, an interim God, a useful temporary God, a makeshift deity, a God who is only a means to something else that is better, less incomplete and less unstable. The ultimacy of theism as well as its range compels philosophers to investigate it. Theism *has* to be metaphysical.

That range, I have also asserted, must be at least cosmological. God may be *over*-worldly; theism may have to treat of many *other*-worldly matters; but theism dare not *neglect* the world. It dare not, as Augustine once said, concern itself wholly with God and the human soul, leaving the world out of account. Therefore, I submit, any philosophical theory that regards either God's earth or his heaven as delectable valleys of culture, temporarily sheltered but precarious and in all probability ephemeral is not theistic. It may palely resemble theism, and if theism be false it may be all that a sane philosophy can offer in its place. But theism it is not.

For that as well as for other reasons it appears to me that neither philosophers nor theologians can afford to be contemptuous of deism. That term, it is true, can be variously construed. Indeed, different authors, preoccupied with their own favourite distinctions, have defined it in a way to which other authors would be strenuously opposed. Kant, for example, defined the term in a very unusual way when, in the *Critique of Pure Reason*, in the section entitled "Critique of all Theology", he said that a deist accepted "transcendental" theology only, whereas a theist also accepted the possibility of "natural" theology, i.e. of arguments based upon the analogy of nature. In the more usual sense of the term (and in the sense in which I am about to employ it) a deist might and commonly did base his belief upon the Argument from Design and upon various "natural" analogies.

For the most part, as we have seen, deism, as contrasted with a more generous type of theism, is said by its opponents to be formal, frigid and thin, if, indeed, it is not more severely treated as is implied in the title of one of Mersenne's books (1624), *L'Impiété des déistes, athées et libertins de ce temps*

Concerning Theism

combattus. A deist asserts the existence of God, that is, of a supreme being on whom the world and its order depend. Beyond that conclusion he refuses to budge. He is usually a monotheist, presumably because it is simpler to suppose one creator than many. Like other enlightened persons he has a great respect for Occam's razor. He admits that the world is not self-dependent, and therefore usually asserts that God was its creator, that being one possible interpretation of the world's lack of self-dependence. Given a wider interpretation a more tepid description of deity might suffice.

A deity so conceived may certainly be conceived very inadequately. He may be too slick to be true. The myth of the "God in the watch-tower" in Plato's *Politicus*[1] may not be very impressive. That deity took the helm periodically to prevent the ruin of the world. Leibniz may have been justified when he rallied the Newtonians for holding similar views. In his first letter to Clarke, Leibniz said: "Mr. Newton and his followers have a very odd opinion about the work of God. They think that God has to wind up his watch from time to time. Otherwise it would cease to go. He had not enough foresight to put it into perpetual motion." Pascal, again, was very bitter about the *chiquenaude* theory of deity in Descartes's physics. "I cannot forgive Descartes", he said. "He would have liked, in his whole philosophy, to be able to be rid of God, but he could not avoid making God give the world a twirl (*chiquenaude*) to put it into motion. After that he had nothing more to do with God."[2]

That was also the opinion in the line regiments of the Church militant during the eighteenth century. Deism was regarded as a device for repudiating true religion without absolutely denying the existence of God. At its best, these critics averred, it was Laodicean. At its worst it was still more unpalatable. There are, however, good historical reasons for holding that very many, if not most of the deists in the eighteenth century were quite sincere in their beliefs, although, on principle, they

[1] 272e.   [2] *Pensées*, 77.

Theism and Cosmology

took the matter pretty calmly. What they wanted was the candid acknowledgment by enlightened common sense of the reasonable element in religion, shorn of vulgar enthusiasm. Their theology was much more than a formal salutation of divinity, although, no doubt, it was comfortable, cool and safe, a belief that allowed classical antiquity to join hands with the modern world, and repudiated ecclesiasticism, fanaticism and the zealotry of special revelation.

Most good rationalists of the Enlightenment, therefore, were willing and ready to be not much more than deistical theists, and the majority of them (with the express exception of d'Holbach and a few of his friends in Paris) regarded out-and-out atheism as emotional and rather indecent, as well as unnecessary and improbable. Because he was a deist (according to several historians) Adam Smith hesitated and finally declined to fulfil his duty as Hume's literary executor. He would not publish the celebrated *Dialogues Concerning Natural Religion*. Smith was shocked at the idea that respectable deistical theism, if treated quite calmly in the detachment of a truly enlightened spirit, was suspect in a serious way and perhaps not greatly superior to the unclassical, evangelical, superstitious Christian zealotry that men of sense believed themselves to have wholly outgrown.

In any case it is necessary, in common equity, to distinguish between what deists asserted, and what they denied either expressly or covertly. If they were wrong in what they denied, they might be right in what they asserted; and their reluctance to go beyond certain simple and thinnish assertions need not have been baseless even if its justification was insufficient.

What deists asserted, then (for the most part with great confidence) was that a reasonable man ought to believe in the existence of an intelligent author of the world-process. That was affirming a great deal. It was challenged, rather than supported, by what appears, enigmatically enough, to be the upshot of Hume's argument in the *Dialogues*, Philo's pensive admission that "the cause or causes of order in the universe probably bear some remote analogy to human intelligence", although a certain melancholy at the obscurity of this thesis was also appropriate.[1]

This positive, although muted, affirmation of Hume's Philo, however, was itself contrasted with the trenchancy of that same Philo's denial that "the cause or causes of order in the universe" could have any analogy whatsoever to "the other qualities of the mind". Philo admitted, in short, that there was some appearance of strength, perhaps that there *was* strength in the arguments for the deist's God, but he affirmed that there was no strength at all in the arguments in favour of the existence of a deity who was *more* than an intelligent world-orderer. This negative thesis pertained to the deists' creed. The deists denied that it was possible to show by any reasoned argument that God was gracious, and kind, and loving. According to their philosophy, the moral attributes of God, as well as his political and aesthetic attributes, were not even problematical. The existence of a problem implies a certain basis for argument. The deists held that there was no such basis. The part of Christian and of some other theisms, therefore, that asserted God's bountiful providence and righteous aims was, in their opinion, simply a piece of unphilosophical anthropomorphism. They were like the men "settled on their lees" in the Book of Zephaniah (i. 12), who said in their heart "The Lord will not do good, neither will he do evil".

If it be complained that deism itself is anthropomorphic, and all too human in its way of thinking about these matters, the answer is that although many historical deists may have earned this reproach, their theory, nevertheless, could be freed, rather easily, from the cruder entanglements of anthropomorphism. The deists believed, it is true, in the unity of divine intelligence, and thought of a human mind as the nearest approach to a unified intelligence that experience offered. The analogy, however, need not have been pressed. It is comparatively easy to unthink such anthropomorphism, and to

[1] *Dialogues*, Part XII, Clarendon Press edition, p. 281.

speak, more circumspectly, of the "unity of consciousness in general" or of a logical unity without any living mind to grasp it. If such ideas fail in the end, they can at least postpone the hour of their surrender for a very long war. In short, a candid deist might reasonably believe that he was not very seriously committed even to the semblance of anthropomorphism. He knows that the world-producer must be very unlike any other producer, and therefore must be very un-manlike in a host of ways. He could admit the dissimilarity without inferring (as Hume's Philo perhaps inferred) that the similarity was almost entirely nebulous.

Certainly it may be argued that a creative God, in the more usual sense of creation, is volitional as well as intelligent, that mere intelligence may be as much of an abstraction and as little of a concrete reality in heaven as it is on earth, and that there would be a shocking lack of intelligence if an orderly world were created for any other purpose than the bringing of beauty or of moral excellence or of some other great value into existence. If so, it might very justly be inferred that the philosophy of a bolder and more opulent theism is more readily defensible than the precarious because over-cautious doctrines of a spare, unadventurous deism. From this point of view the appearance of anthropomorphism may be simply another way of saying that man, in the finer part of him, is made in God's image and a little lower than the angels, or, more modestly, that God is less inadequately portrayed when he is represented as akin to man's nobler attributes than when all such resemblance is denied. Anthropomorphism itself, some would say, is defensible, and theism is the stronger for employing such conceptions.

However that may be, it is preposterous to assert either that mere deism is the sum of "natural" theology, or that there can be no philosophical grounds for a theism that exceeds mere deism. My plea for the recognition of deism in natural theology is not conceived in any such spirit, and I am anxious to say that it is not. On the other hand, the deist's caution, even if it

Concerning Theism

is mistaken, should not be summarily condemned. His reluctance to wander blithely and almost carelessly on the resilient turf of familiar Christian apologetics is reasoned and may not be unreasonable. Deists may be illiberal and short-sighted in their views, and their opponents may be better cosmologists than they are; but even a narrow cosmology may be better than a perfunctory cosmology. If cosmology is an embarrassment to theism, such a theism is self-condemned.

Deism, to be brief, is a species of philosophical theism. It is a cosmological theory of the origin and stability of the universe, conceived upon principles that philosophical theism may rightly decide to incorporate. It is bad policy on the part of a theist to neglect what deists assert through distaste for what mere deists deny. If deists draw their boundaries in the wrong places, their mistake, as regards the philosophy of it, has to be shown by philosophical argument to be a mistake. It has also to be shown that there are no similar boundaries in any other place, for example, regarding God's "personality"; and the notion that there are such boundaries, to certain apologists, is an idea almost as disasteful as mere deism itself. Frontiers are obscured but not removed by raising a dust in their neighbourhood.

I hope that these rather lengthy discussions about the meaning of theism, pantheism and deism may be of some service as an indication of our theme as well as of the words that are to be used in describing it. In any case the time has come to make an end of preliminaries, except in one essential respect. That exception is the order of exposition I propose to follow. I shall now attempt to indicate it.

I am calling the first series of these lectures "Theism and Cosmology", and I shall call the second series "Mind and Deity" This selection of titles, as I know very well, may generate misunderstanding. I do not mean to be rid of cosmology at the end of the first series, and to ignore it in the second. On the contrary, if, as I have said, philosophical theism must at least be cosmological, that requirement pervades the entire

Theism and Cosmology

treatment of the subject. Again I do not mean that cosmology should be discussed without any reference to what may be supposed to be the marks of a governing mind in an ordered universe. That would be a travesty of most theistic cosmologies, including the most deistical among them. I shall explain in a moment what I do mean with regard to this matter.

I intend in the first series to examine a part and, largely, the deistical part of cosmological theism, that is to say, the cosmological argument itself, the relation of "first" to "second" causes in nature and the need for the former, the problem of origins in theories of creation, emanation and the like, the theological implications of space and time, the evidence that the order of nature must be planned. These problems form a relatively compact group, and therefore I shall treat them together in the first series of these lectures.

In that series, I shall also treat these questions in a provisionally realistic way. That is the sense in which most of these arguments have been usually understood, and it is their obvious purport. Such a provisional realism should not be objectionably question-begging, although it provisionally makes its humble plea in the knowledge that it is impossible to discuss everything all at once. By the provisional realism of this discussion I mean the assumption that non-mental entities may exist and come within the ambit of human knowledge. In other words, I mean to discuss cosmological questions in the first series without considering the arguments in favour of what is usually called an idealistic philosophy, that is to say, I mean provisionally to neglect the implications for cosmology of the philosophical view that either all reality or all knowable reality can be shown to consist either of minds or of their ideas. Similarly I shall neglect such views as the ancient comment of Sallustius, "The universe itself can be called a myth, since bodies and natural objects are affirmed in it, while souls and intellects are concealed".[1]

In the second series of these lectures, I shall begin by dis-

[1] Nock, *Conversion*, p. 243.

Concerning Theism

cussing these very idealistic arguments in order, among other things, to discover, if possible, whether the provisional realism of the first series was a serious menace to the strength of the arguments that were there examined. That is one of the purposes of the fresh start I shall make in the second series; but, of course, it is not the only purpose. There may be, and I think there are, theistic arguments that would carry very little conviction to a philosophical realist, but would properly carry great weight with many philosophical idealists. The famous Ontological Proof may be one of these; and that is one of the reasons why I shall postpone discussion of it until we come to the second series.

More generally I mean to postpone to the second series all theistic arguments that, in an obvious way, would seem to be greatly strengthened by the philosophical premiss that nothing non-mental or unspiritual could possibly exist. I say "in an obvious way" because it is quite possible that theism in the end is not more effectively defensible by an idealistic than by a realistic philosophy, and even that some types of philosophical idealism are definitely anti-theistic in a way that few if any realistic philosophies are compelled to be. That also is one of the questions I want to discuss in the second series of these lectures.

Despite the intricacy of this question, however, I incline towards the belief that it is better as well as more convenient to discuss the part of theism that philosophers like Mr. Taylor would oppose to mere deism after and not before an attempt has been made to consider the general view that whatever exists is a spirit or the ideas of a spirit.

God's moral providence, his tenderness, grace and pity seem to me to be most appropriately discussed at this later stage of the enquiry and not at any earlier stage. I do not, of course, deny that nature, realistically understood, may supply evidence of a just and compassionate as well as of an orderly and intelligent governing mind. The bounty as well as the orderliness of the cosmos may be evidence that supports or compels the theist's conclusion. *Prima facie*, however, care and tenderness and justice seem to be properties that are most readily intelligible in a companionship of spirits, infinite as the inequalities among such spirits may be. Therefore if they pertain to the world-order it is easier to assent to such a view if the world-order be in fact an order of spirits than if it is largely non-mental in a way that may seem to threaten the destiny, security and felicity of spiritual, and in particular of moral beings. That *prima facie* case is sufficiently strong, I think, to make the course I intend to follow not quite arbitrary. I do not claim a greater justification, but I do not think that I need one.

III

The Cosmological Argument

THE term "the cosmological argument" is the traditional name, in Kant's and in some other terminologies, for the argument "a contingentia mundi", that is to say, from the self-insufficiency of the world, where the "world" is understood to mean, not the Earth, but what Christian theologians would call the entire created universe. In Greek philosophy the argument was from astronomical motion to the *Primum mobile*. A wider form of a similar argument would be the attempted proof of theism from the self-insufficiency of ostensibly empirical beings. As Proclus said, "All things pray". The latter type of argument, however, need not be identical with the former, since it is clear that the entire compages of finite things might be self-sufficient, although no particular finite thing was self-sufficient. I shall, however, consider both of these types of argument in the present lecture.

Such arguments express the empirical approach to theism in the most general way in which that approach can be expressed. They are, for instance, enormously wider than the familiar argument from design, which plunges headlong into a special interpretation of the facts. Consequently they are apt to seem very abstruse. It is the middle axioms of philosophy, not its austerer first principles, that are usually most persuasive. Philosophy, however, is the jealous guardian of the most general principles, if haply it can find them, and the cosmological type of argument seems to be a better logical introduction to empirically grounded cosmological theism than any other. Even if, strictly interpreted, such arguments are only a step towards theism, and never the end of the theist's proof,

Theism and Cosmology

it is very important to enquire whether such a step must be taken. However scornful modern theists may profess to be about the traditional metaphysical proofs of divine existence, they commonly deny in fact that either the universe or anything in it is self-sufficient; and they use the denial in support of their theism. The question should therefore be investigated methodically. The discussion may be dry: but if I could de-bamboozle and unflummox myself and some of my hearers I should have done quite a lot.

The logical structure of the argument *a contingentia mundi* consists firstly of the assertion of an existential premiss (viz. that the world exists) and secondly of an inference by complementary relation. Insufficiency, it is claimed, implies a specific complementary sufficiency. We have therefore to examine this logical structure of the argument in order to see what it would prove if it were set forth *secundum artem*, and, by implication, to discover whether its terms are as general as they seem to be or conceal more special and very disputable assumptions.

The fact that an existential premiss is asserted is of the very greatest moment. It is a very general (and, I shall here allow, a very sound) philosophical belief that although existence may be inferred from other existents, no existent whatsoever can be inferred from mere *conceptions*. Therefore if anyone proposes to prove that God exists, he must anchor his proof to what is admitted to exist. If he does not do so he runs the risk of arguing to some God-like idea or ideal, instead of attempting to prove that there is a God.

Exception may indeed be taken to the existential premiss that is asserted in the argument *a contingentia mundi*; for the "world" is an elaborate thing, and so it may be objected that the proof would be very much stronger if it rested upon a simpler existential premiss. Such a premiss would be that certain things exist, or that the philosopher who uses the argument exists when he uses it, or, simplest of all in appearance at least, that something exists. A prudent philosopher it is claimed would attempt to rely upon the *minimum irrefragabile*

The Cosmological Argument

in this kind rather than to build upon a more complicated but more disputable existential premiss. The proof that, if anything exists, God must exist would therefore be stronger than the proof that God must exist because the world exists, and so would be preferable if it could legitimately be drawn. For it is desperately hard to deny that something exists.

I shall return to such topics very soon. For the moment I am content to call attention to the presence of an existential premiss in the argument and shall next deal briefly with the inference by complementary relation.

This type of inference would seem to be valid. If a man exists who is married, a woman must exist who is his wife. If a male exists who is a brother, either his brother or his sister must also exist. The latter example, however, shows very clearly how careful we must be to determine precisely what the complementary relation is when we draw such inferences. A brother need not have a brother; for a sister might do. It is further to be remarked that many such inferences can only be drawn if there is independent evidence of the very thing that is to be proved. To prove that such and such a man is married is to prove his marriage to such and such a woman. To prove that X has a brother or sister is to prove that his parents had more than one child. In the case of theism we have therefore to be very careful indeed to ascertain whether the relevant inference by complementary relation could be drawn, unless there were other evidence that established it already.

I said in the last lecture that theism was not seriously philosophical unless it asserted the existence of a divine being, singular or composite, who is ultimate, a unity, and operates upon a cosmic or hyper-cosmic scale. We have now to ask whether an inference by complementary relation could yield *such* a complement, or point clearly towards such a complement.

It seems unlikely that any lesser premiss than "the world exists" would be sufficient for the argument. The argument, in other words, would seem to be stronger in its more usual form, that of the *argumentum a contingentia mundi* than, say,

Theism and Cosmology

in Locke's form[1]: "I exist; therefore God exists", or in the still more importunate form: "Something exists; therefore God exists".

This remark may not be more than a tautology. To argue from the contingency of the world, "nature", the cosmos or the universe, is, of itself, to argue upon a cosmic scale. Therefore it is an argument *to* a cosmic ground. The other arguments that proceed from a narrower base would require an additional premiss in order to show that the complement they inferred was of a cosmic or hyper-cosmic order. That premiss would have to be an assertion about the *mundus* itself.

Take, for example, the premiss "I exist and am not self-dependent" and allow, for argument's sake, that an existence that is not self-dependent implies an existence that *is* self-dependent. My lack of self-dependence shows that I am not alone in the world. I have had parents, and they and their ancestors, by luck or good sense, or both, must have had a good deal of vitality. On these lines, assuming the validity of ordinary causal inferences, one can infer a good deal, namely the existence of all the cause-factors that have been relevant to my birth and subsequent career. This cause-cluster, however, is but a tiny part of the universe, and the entities on which I depend, or have depended, are of the same order as I am myself. Nothing in the argument shows that any of them operates on a cosmic scale.

Therefore Locke's argument (which I choose because it is typical of its kind, and not because it was worthy of Locke's genius) had rapidly to introduce the cosmic premiss. From the disputable principle that a cogitative effect implies a cogitative cause, Locke inferred[2] that there must be a single cogitative First Cause which must have produced all the "order, harmony and beauty which are to be found in nature". All that his reasoning entitled him to infer, however, was that, since he existed, there must have been a sequence of cause-factors sufficient to produce all the order, harmony and beauty, as

[1] *Essay*, Book IV, chap. x. [2] Ibid., § 10.

The Cosmological Argument

well as all the disorder, ugliness and discord, that were to be found in *him*.

Totalitarian philosophers, it is true, usually claim that from the premiss "This exists" (whatever "this" may be) it is always possible to reach the Whole by a logical process. Hence, if the Whole were divine, totalitarian pantheism would follow from "This exists". I submit that there is no such logical short-cut to the Whole. There are no known logical or causal principles according to which the toad-flax on a derelict garden-wall supplies sufficient data for inferring everything else that there is, for example, the weight of the world's oldest white elephant. If it be said that by tracing the history of wind-borne seed, and the causal ramifications of each event in that history, and so on, nothing in the end would be left out, the answer is that it would still be incredible that the toad-flax and its seed contained specific marks of the All. The totalitarian makes his journey *via* the world. He contends that one finite thing leads to another, and this to a third, until the whole of nature is included. From nature he may advance to pantheism, but he does not directly proceed from a bare *something*, from anything, to the All.

Therefore I think this argument is most effective when it really is cosmological, that is to say when the existence of the cosmos is its existential premiss. In that case it is still very general. It does not, like the argument from Design, argue from any specific characteristics of or within the cosmos to a hyper-cosmic source. On the contrary, it argues quite generally from the debility of the cosmos at its proudest. On the other hand, thus interpreted, it cannot really be a very simple argument. The debility of the cosmos at its proudest would have to be established, and then it would be necessary to show what sort of debility there was, and what sort of remedy this debility not only demanded but also, metaphysically speaking, was bound to receive.

Accordingly the operative existential premiss in the cosmological argument is that the world, the cosmos, the universe exists.

Theism and Cosmology

I do not suggest that this premiss should be glibly asserted without sedulous examination. On the contrary, the man who says that he knows that the universe exists should be prepared to undergo a gruelling cross-examination. Does he know that the cosmos exists just as he knows that he is now making (or hearing) a sound when he says so? What does he mean by the universe? Is it the physical universe? If so, what does it include and what, if anything, does it exclude? Does he or does he not covertly assume that "the universe" is just one thing, that it is a single whole, finite or infinite, that it is integrated in a single pattern, and the like?

Such questions cannot be avoided and I shall return to some of them in the present lecture, constrained by another turn of the argument. For the moment, however, I want to allow, provisionally, that there is some intelligible sense in which we can assert that there is a "cosmos", and I want now to attempt a closer scrutiny of the second step in the cosmological argument, viz., its inference by complementary relation.

In general terms, the argument is an argument from the self-insufficiency of the cosmos. What we now want to know is the type of self-insufficiency that is supposed to demand a theistic complement. As the word "contingency" is used in the official traditional designation of the argument we may conveniently begin with it. What is the contingency of the world, and why does such contingency imply one and only one complementary? Why, further, must this solitary complementary be God?

Unfortunately the term "contingency" as employed in the traditional expositions of the argument had several meanings, and was consequently ambiguous when its multivalence was not explicitly remembered. Therefore Kant had justification for his complaint that the argument, as developed in the Leibnizian tradition with which he was familiar, "contained a perfect nest of dialectical assumptions".[1] The critical philo-

[1] *Critique of Pure Reason*, Section on the Cosmological Proof.

The Cosmological Argument

sophy, he said, had to destroy the nest. In any case the multivalence of the term has to be explored.

(*a*) One of the more ordinary meanings of the word "contingent" is "fortuitous", but it seems plain that if this were the relevant meaning of contingency, the cosmological argument would be *very* badly expressed. Theists, it is true, frequently argue that the world, left to itself, would be a haphazard affair, but that, since it is not haphazard, it is not left to itself. Such an argument is at least intelligible, but it can hardly be reconciled with the usual form of the cosmological argument except by a quite prodigious piece of prestidigitation. It is not an argument from the randomness of the world (for that is denied) but an argument from the assertion that the world would be a chance affair if (as is not the case) it were left to itself. Instead of a descriptive existential premiss we should have a presumptive description of what does not exist, and the argument from complementary relation would be only a sham. There would be no argument from chance to what was not chance, but only an explication, with rather unnecessary circumlocution, of what is supposed to be known already, namely that nothing does occur at random.

If the argument were from the presumptive randomness of the world *if left to itself*, we could certainly ask *what* complementary factor annuls the randomness[1] that *would* occur in a world left to itself, but that does not *in fact* occur in the world or outside it. That appears to be a decisive question, but there would seem to be several candidates for the position of complementary opposite.

(1) In one sense, chance is opposed to design. We speak, for instance, of a "chance" or of an "accidental" meeting between friends, or acquaintances or enemies who had not intended to meet. If so, and if the cosmological argument could employ the conception, it would be a short-cut to the

[1] Or "chaos"? But chaos is not *chance*, e.g. if a bomb reduced Glasgow University to chaos by explosive causes.

Theism and Cosmology

Argument from Design. We shall discuss that argument later, and shall find that there are no short-cuts to it.

(2) Another sense of "chance" or "accident" may be illustrated by the "accidental" collisions that are usual among the toy boats that are sailed in the smaller ponds of a public park. This means that when the separate causal histories of the individual boats are alone considered there are occasional violent interruptions from what are regarded as external and independent causal histories. It is not meant that the collisions occur without any cause or reason.

The complement of "chance", in this sense, would be an external and independent causal series. Suitably modified, the cosmological argument might then state that the causal sequences that should be presumed if nature were left to herself imply non-natural causes of *some* kind if the actual course of nature (which is not left to herself) can be made intelligible.

(3) In yet another sense, very often employed in the fashionable theories that attempt to reduce probability-statements to statements of class-frequency, a "random" distribution is a certain type of statistical distribution. Here the associations of the term "randomness" seem still to be what logicians would call "privative", that is to say they denote the absence of what might reasonably be expected to occur. Such associations may be misleading, and I think they are misleading in many modern discussions of "randomness". It is, for instance, quite illegitimate to argue that because certain aggregates appear to be "random" in the sense that they do not indicate the pervasive presence of certain assignable cause-factors, therefore the members of these aggregates are not causally determined.

Nevertheless it is plausible to maintain that the name "randomness" may be and should be understood in a spirit of acquired innocence. A "random" distribution, we may be told, is simply one type of distribution. It is the sort of distribution that shows no trace of the sort of pattern or heaping that would

The Cosmological Argument

be apparent if this or the other cause-factor were probably operative. It shows no traces of any particular cause-factor. The absence of such a cause-factor might, of course, be simulated, as when an "accident" is staged; but a possibility of that kind is not an objection to the contention.

On these assumptions the relevant contrast would be between the pattern of a "random" distribution, on the one hand, and, on the other, any sort of *specific* causal pattern. It would not appear that such conceptions offer a hopeful approach to theism by way of the cosmological argument. The existential premiss would appear to be the presumption that nature, left to herself, would exhibit a "random" distribution and that only. Such a presumption, on merely general grounds, would appear to be quite arbitrary. Indeed there is no known theory of probability according to which apparently "causal" distributions should *not* be expected to occur in a very large "random" distribution. There might well be pockets of stability in a welter of chaos. Passing the point, however, we should have further to remark that the "causal" alternatives to "randomness" would be legion. Hence it is very unlikely that the only correlative to presumptive diacosmic "randomness" would be cosmological theistic causation.

That illustrates the danger of arguing from privative terms. In using them, we are far too apt to assume that we know in advance the precise character of the deprivation. Thus blindness as applied to a man is supposed to be a privative term, because it is supposed that a man "ought" to be able to see. It would not be a privative term if it were applied to a mole at the present stage of the evolution of that species. Plainly, however, we have no right to assume, or to presume, that blind men cannot exist, and we have no right to argue from general debility to any particular kind of complementary strength, even if we have the right to argue that sight alone could remedy the particular deprivation called blindness.

(*b*) Let us pass, however, to another and quite different sense of contingency, namely dependence. This is the sense in which we say that something or other is contingent or conditional upon something else. Such "contingency" is obviously quite different from randomness. If, as Spinoza held, finite existence followed *ex necessitate divinae naturae* it would be contingent in the sense we are now about to examine, but would be necessitated and not at random.

There are two principal types of dependence and these are, firstly, the dependence of a logical consequence, and, secondly, the dependence of an effect. The two may be connected in so far as a "cause" is also a "reason", and the word "ground" is sometimes employed to express both logical and causal antecedence. It seems plain, however (as already observed), that a non-existent ground could not generate an existent consequence. In the cosmological argument, therefore, the critical question is the derivation of an existent from an existent, and the theistic proof of the argument is *a posteriori* because its inference is regressive in that sense.

In its usual form the argument runs: "The world exists, but it is an effect. It must therefore have a cause other than itself, and the name for this ultimate extra-mundane cause is God."

That is the usual argument to a first cause, and we shall encounter it very frequently in subsequent lectures. The same, however, will be true of most of our discussions in this first series of lectures, for the cosmological argument states in general terms the sort of thesis that may be elaborated in greater detail and supplemented in specific ways. What is attempted in the present lecture is a preliminary survey of the logical structure of most of the subsequent theistic proofs. So let us go on with the argument.

If the world is an effect, and if it can be shown to be so, it must have a cause or set of causes. Again, on these assumptions, if it does not cause itself, its cause or set of causes must be extra-mundane. Moreover this cause, or set of causes, since the world is their effect, must operate upon a cosmic scale.

So stated, the argument would not prove that there is a *first* cause, but it would prove the existence of an over-worldly

The Cosmological Argument

cause or set of causes operating upon a cosmic scale. *Prima facie* at least there would be no difficulty about the inference by complementary relation from the existential premiss. For, plainly, the complementary term to "effect" is "cause". On the other hand, it would be very difficult indeed to prove the truth of the several propositions contained in this statement of the argument, and it is also very difficult to believe that if these several propositions could be proved, the cosmological argument would be other than superfluous, its conclusion having already been reached. I shall call attention to these matters rather than to the special difficulties regarding a "first" cause, reserving the latter for subsequent treatment. If the cosmological argument could pave the way to theism without completing its road-making it would have accomplished a great deal, and we may well be content, for the present, to examine it in that spirit.

Take, first, the statement that the world is an effect. If so, it must have a cause or set of causes, but how could we show that it is an effect? Might it not be the theatre in which all causes and all effects occur, but itself neither cause nor effect? Are we not driven to say this about *something*, say of "reality"? If so how can we prove, in general, that "reality" and "the world" are not identical? Even if everything in the world were an effect, as determinists hold, the world itself need not be an effect.

Again, many philosophers have maintained that it is not true that everything that exists, or even that everything that has a beginning, has a cause, that is to say, is an effect. The world, they say, contains "spontaneous", free, or uncaused and unoriginated events. In any case they assert very positively that there is no way of proving that such uncaused events do not occur. This question, no doubt, is still the subject of strenuous philosophical debate, but at least it is plain that the proposition "The world is an effect" is most highly disputable and that the proof of it, if there is one, has been so frequently mishandled by philosophers of the highest standing that it cannot be classed among the easier verities. If, then, it cannot be (or has not been) shown that every event must have a cause, how could it be shown that the world must have a cause? Might it not belong to the company of the uncaused? Would it be possible to persuade anyone into the belief that the world had a cause although many of the events within the world had no causes at all?

Once again, how could it be shown that anything is an effect unless it is, firstly, the later term in a sequence, and secondly is tightened up into the sort of sequence that we call an instance of *propter hoc* and not merely of *post hoc*? We surely cannot pretend that we know that the world is the later term in a sequence, for we have no experience of anything that came before the world. Besides, it is very generally maintained by philosophers that our criterion of the distinction between a causal and a casual sequence depends upon repetition of the sequence in varying circumstances, no negative instances having been discovered. Since the world is unique it need not resemble such recurrent and similar sequences *within it* in any relevant way. Therefore the ordinary means for discovering a cause or an effect would seem to fail us altogether, and if, by what most philosophers would consider a miracle, we could see that the world was an effect, the cosmological argument would be a miserable tautology since the premiss that the world was an effect would just be another way of saying that it had a cause.

Let us, however, consider certain possible variations in the argument from cosmic dependence. In my statement of it, I have included the qualifying clause "if the world does not cause itself". This qualifying clause may be held to be redundant on the ground that the very conception of a self-causing cause is inherently absurd. Cause and effect, it may be said, must always be different. Therefore nothing can literally cause itself. That is obvious if a cause must be an antecedent. Nothing can antecede itself. If cause and effect could be simultaneous, that particular argument, it is true, would not

The Cosmological Argument

apply; but it might still very reasonably be held that even if certain causes were simultaneous with their effects, they must nevertheless be different from their effects.

So far, the cosmological argument in the form in which I have stated it, would seem to be strengthened. It would now run: "The world is an effect, and nothing can bring itself into being. Therefore the world must be the effect of *another* existent operating upon a cosmic scale." This apparent advantage, however, might rapidly turn into a heavy liability. If nothing can cause itself, the ultimate ground of the world cannot cause itself, and so, if it exists, it must be an uncaused thing. Therefore something must be uncaused. A theist would say that this uncaused being is God, but his opponents might submit that the world was a promising candidate for the same position.

I need not, I think, do more than mention the sophism that misled Locke, Clarke and many other eminent men. Hume[1] has done that with finality. Theirs was the hollow argument that if the world were uncaused it must have been caused by nothing, and that non-entities do not act. It is enough to refer to Lewis Carroll's *Through the Looking Glass*. " 'I see nobody on the road', said Alice. 'I only wish *I* had such eyes', the King remarked in a fretful tone. 'To be able to see Nobody! And at that distance too!' "

These considerations, I think, are sufficient to show that if the alleged "contingency" of the world means its causal dependence, the cosmological argument is speculatively insecure. A direct proof of such dependence would have to be supplied. If it were supplied the cosmological argument would become a trivial exercise in formal logic. The same remarks would apply to the contention that might seem to be but (I think) is not more general, viz. that the world is the logical consequence of another *existent*.

For better measure, however, I shall add some other observations. To say that the world is not self-dependent is consistent with the view that a *slight* extra-mundane cause-

[1] *Treatise*, Book I, Part III, § iii.

Theism and Cosmology

factor might complete it. Theists make very much bolder claims. Again, the world's dependence might be taken to mean that it was an assemblage of dependent entities. In that case it has been argued that the world might depend upon an assemblage of causes, no one of which need be different in type from its specific effect, and all of which, taken together, might be as little ultimate and as little unified as the assemblage of effects that we call the world. Suppose, however, that the world had but one cause. That cause might be an angel working under an archangel, working under a demiurge, working under ... I allow that, for the time being, I am considering the cosmological argument rather as a possible first step towards theism than as a final demonstration of theism. That circumstance may diminish the force of the objection. There is some importance in noticing, however, how very short this first step would be, supposing that it were a genuine step, and not simply the raising of one foot in the air.

(c) A third meaning of "contingency" is the sense in which the "contingent" is opposed to the "necessary", and the cosmological argument frequently took the form of maintaining that since the world was a contingent being, and existed, it implied the existence of a necessary being. The nerve of the inference was that "necessary" is the only relevant logical complementary to "contingent".

If so, strong exception should be taken to the argument. I shall begin by considering the most powerful counter-argument, although it is rather abstruse.

It is, in substance, that "necessity", to use traditional logical terms, is a modal predicate, and is not, as the above argument fallaciously assumes, a descriptive or assertoric predicate. Modal predicates, to put the point in other language, do not apply directly to things or to what is existent, but apply only to propositions. They are secondary or second-order propositions about primary propositions, and neglect of this crucial consideration leads to a *sophisma figurae dictionis*, a fallacy of figure of speech.

The Cosmological Argument

Expressed at greater length, the objection is that modal assertions cannot be part of any primary proposition but belong to a certain group of secondary propositions. That is the group that describes a certain type of relation between the subject and the predicate of a primary proposition. The relation in question is the relation of being certifiable or uncertifiable by inference, and the truth of the matter is that the genuine subject of such a second-order proposition must always be an entire primary proposition. It is never legitimate to say that "S is necessary" in the sense in which it may be legitimate to say that "S is P". All that can legitimately be said when the modal predicate "necessary" is used is the statement that the entire proposition "S is P" is necessary. In the instance of theism the appropriate comment would be that we are not entitled to say that God is necessary (or is a necessary being) in the sense in which we might be entitled to say that God is omnipotent (or is an omnipotent being).

I have stated this objection in the traditional logical terms of subject and predicate, but there would be no difficulty in expressing it in terms of what would be currently regarded as a more modern form of formal logic. It applies in all such theories to the term "necessary" and also to the term "contingent" if the latter term be defined as the logical correlative of "necessary". If the one is modal its correlative is also modal. So the pertinence of the contention to the traditional cosmological argument is very obvious indeed. There is no such thing as a contingent being, the objector says, and no such thing as a necessary being. Therefore the argument that the world is a contingent being, and consequently implies a necessary being, owes all the plausibility it may possess to a fallacious way of speaking.

I think it is clear that these objectors are in the right, and I might add that although the path by which we have reached this adverse conclusion is stony and hard, the conclusion itself has a comforting and even a homely look. There does seem to be something wrong about the conception of a "necessary being". The phrase would appear to mean "a being that necessitates itself", and necessity, it seems plain, applies to arguments and not to things. Nothing, not even God or the world, can argue itself into existence, and if we were to say that a necessary being is one that necessitates itself, we should have returned, in an awkward and troubled way, to quite another sense of necessitation, namely causal determination by an ultimate self-causing cause. I do not propose to repeat the observations I have already made about this form of the argument.

There is, however, at least one other matter of importance that falls to be considered in the present special connection. What is "necessary", it may be said, is what *must be*; what is contingent is what *need not be*, i.e. the "possible". Granting that objectors to the cosmological argument are right in their assertion that these statements, if accurately expressed, should refer directly not to things but to propositions about things, not to what must be and to what need not be, but to what must be *such and such*, and to what need not be *such and such*, the fact remains that the proper correlatives for an argument by complementary relation are the "necessary" and the "possible" respectively.

This suggestion, supposing it to be tenable, is itself somewhat perplexing, for the existential premiss in the cosmological argument is not that the world is possible, but that it is actual. One of the reasons for the perplexity is that the traditional list of modal predicates, possibility, necessity and actuality, seems itself to be confused. There is not a triadic correlation, but a cross-division, and the truth (if I may be pardoned for making a blunt observation upon an intricate matter) would seem to be that in all cases we should speak about the necessity or the possibility of propositions *that apply to actuality*. Passing the point, however, we may still be asked to allow that there is a sense in which the existential premiss of the cosmological argument, although it asserts actuality, also asserts an actuality that is merely possible. What it asserts, we may be told, is that

The Cosmological Argument

although the world happens to exist, it might or might not have existed. Therefore although it does exist, it is, metaphysically speaking, only a "possible existent" when regard is paid to its own proper status. We can say of it "Yes. There it is. But there might have been no such thing. There would be no contradiction, no metaphysical absurdity, if something else had existed instead of it, or if there had been no world at all."

Among the many problems that such an argument raises it may here be sufficient to mention only one. Suppose it were true that we could say of something (whether or not that something be the world) that it might or might not have existed, that it is simple or brute fact with no self-explanatory resources. What then? May not all explanations refer to the existent, to the actual? If so what right can anyone have to the assertion that existence must be self-explanatory, able to give a reason why it should be thus and thus, not something else, able to prove that there would be some sort of internal contradiction if that which occurred had not occurred? To say so is not to deny that the existence of X may supply a reason for the existence of Y. Parents do have children. What is denied is that existence can explain itself in the sense that it can produce its own reality in any logical way, instead of being simply *that* to which all explanation must refer.

If this contention be sound, as I think it is, the cosmological argument, so interpreted, melts away like snow in spring. Its existential premiss in this form of it would be that although the world is actual it might not have existed. That statement, according to this criticism, would apply to all existence, except in the special case in which one existent generates another in the way (if any such way there be) in which the progeny can be inferred with logical necessity from the parent. On these grounds alone there would be no inference to an existence complementary to the existence of world. Such a complement would be an existence that from its own internal necessity could not have been different from what it is. The critics assert that although everything is what it is, there can never be a logical contradiction in its non-existence. Deny it, and you deny its self-explanatory potency.

(*d*) At the beginning of the present lecture, I said that the cosmological argument was an argument from the self-insufficiency of the world. Such alleged self-insufficiency includes what is often called "contingency", and I have discussed some of the principal senses of the word contingency that seemed, or might have seemed, to be relevant to the argument. I have now to ask whether there are any other relevant senses of self-insufficiency without requiring that these should be or are usually designated by the word "contingency."

What would usually be meant by self-insufficiency is incompleteness. Accordingly, we may take the cosmological argument to assert that the world exists but is an incomplete thing, that reality must be complete and so that the world implies the existence of some complementary being or beings. For the purposes of serious philosophical theism it would have further to be shown that the incompleteness of the world was radical and profound, and that the over-worldly complement that was required was something more majestic, by far, than the world. It might, however, turn out to be the case that if the first step had to be taken the second would not be beset by insuperable difficulties.

Let us try, then, to consider some of the principal senses in which it may be held that nature, or the world, or what Christian and Mahommedan theologians call the created universe, (including its physics, its biology and its psychology) is incomplete.

(1) One such sense would be "fragmentary" and that simple interpretation has the signal advantage of allowing the existential premiss in the cosmological argument to be understood quite literally without any *arrière pensée*. For fragments do exist, and are what they are—crumbs of bread, lines of verse, fossil bones. From such fragments, in an empirical way, we

The Cosmological Argument

can draw inferences generally accepted as valid. The existence of the crumb is evidence of the existence of a loaf; the lines of verse may show, from internal evidence, that they came, almost certainly, from the pen of such and such a poet; the fossil remains may be evidence for the sort of reconstruction that contributed to the immense reputation of a Cuvier.

All such empirical arguments, however, depend upon specialized empirical knowledge. They would lead nowhere in particular if all we knew was that the thing before us was *somehow* fragmentary. Therefore they would lead nowhere in particular if all we knew about the world was that it was *somehow* incomplete in this sense. Moreover, such empirical fragmentariness is usually held to point the way towards an empirical completion of the same order. *Ex ossibus megatherium; ex pede Herculem.* Consequently, if it could be shown that the world is a fragment, something like an immense meteorite, very little could be inferred that would be likely to assist theism. Indian philosophers might call the world one of the sixty-one great chiliocosms each of a thousand worlds. If it could be shown that the world is a very special kind of fragment that implied a highly specific missing part (although perhaps a very great missing part) we could not draw the required theistic inference unless we knew what the whole was like. (Cuvier, for instance, had a profound acquaintance with the animal structure of *whole* animals.)

(2) A second interpretation that may be put upon the professed incompleteness of the world is that the world has an incomplete kind of existence, a low "degree of reality" as some philosophers would say, an ontological "diminution of being," as M. Maritain would say.[1] Such statements appear to me invariably to involve an abuse of language. To be is just to be. Whatever is, is. There are not several distinct ways of existing, one of which is full existence and the others some kind of partial existence. What exists does exist. If it is a part, then something else exists, but it is not true that a part can have

---

[1] *The Degrees of Knowledge*, English translation, p. 274.

Theism and Cosmology

only a partial existence. It may be possible to employ such language without very sinister consequences. Grammatically there may be a fairly harmless zeugma which should not deceive. Sylleptic philosophy, however, is always dangerous, and all serious and determined philosophical talk about half-being, partial being, three-quarters or five-eighths being seems to me to be reprehensible, even when it is Plato who does the talking.

Beings may be feeble, transient, trivial. In other words they may exist and yet be weak, be real and yet be ephemeral, be actual and yet be unimportant; but it is always inaccurate to say that they have a weakly, transient, or unimportant "mode of being". Philosophers, as it seems to me, have sometimes cozened themselves into this maze by inventing the monstrous hybrid that nowadays is sometimes called an "epistemological object." The phrase reminds me of a detective story I recently read in which a skeleton key, used by the investigators to effect an unlawful entry, was called a "provisional key". There are (I believe) things, and (in my opinion) there is such a subject as epistemology. But there are no epistemological things. The phrase designates a certain status of things in relation to some apprehending mind. It does not designate a special epistemological sort of thing. There is no epistemological "mode of being", that is to say no entity that exists but only in the epistemological way.

The hybrids that are erroneously supposed to be conjured into existence[1] in this way, are usually held to be compounds of mental and non-mental, things that are "mixed with mind" like dream-castles and "false objectives", and the exploded theories with which a Milton might people his Limbo. If there really are minds as well as non-mental things, and if minds, as Mr. Whitehead would say, may be "ingredient" into an existence beyond themselves, then there might be "mixed" events, in whose constitution mental and non-mental mingle. If so, these "mixed" events exist just like any other actual

[1] Or is it semi-existence?

The Cosmological Argument

mixture; but they cannot have a mixed sort of existence. One might as well refuse to see any difference between detecting and perpetrating an ambiguity.

(3) Accordingly we should dismiss the suggestion that the world exists in an inferior sort of way, with a mixed or partial mode of being. From the standpoint of epistemology what should be said if these portmanteau expressions, like Revelation suit-cases, were duly expanded, would be that when we think of the world we think illusorily. In that case the existential premiss of the cosmological argument would be, not that the world exists, but quite the contrary. It would be that the world is illusory, that is, does not exist. It would be *Maya*, even if it were God's (or Brahman's) *Maya*.

If it be maintained that the existence of an illusion implies the existence of something that is not illusion, the answer, I think, does not readily support the cosmological argument. It is usual to distinguish between mere illusions and grounded illusions. If the world were a *mere* illusion the existence of such an illusion might tell us something about our minds, but need not convey any other information whatsoever. If the world were an illusion, but not a *mere* illusion, then we would know that although the world was in part illusory, it bespoke the reality of something not illusory. This view of the world is of course very common in theological circles. It may be true, and there may be persuasive evidence in its favour; but it does not readily conform to the traditional cosmological argument in favour of theism. From the premiss that the world is a well-grounded illusion no inference follows by complementary relation except that there is a reality misleadingly reported (perhaps among other reports) by what is called the world. What has to be shown in order to obtain the theist's conclusion is that the world is illusory *because it is de-theologized*. If that could be shown there would be no need, and I do not think there would be any place, for the cosmological argument. I allow, in conformity with what I have already said, that a step towards theism, even if the step be very short,

Theism and Cosmology

would be a theistic gain, but I cannot see that the well-grounded illusoriness of the world is such a step. If the step were made, the reality to be inferred might be very like the world and quite undivine. Again, as some philosophers have held, both God and the world might be well-grounded illusions. It seems to me therefore that those who would maintain that the world is illusory *because it is undeified* should avoid the cosmological form of argument.

(4) Certain recent philosophers have recently elaborated a doctrine of "incomplete facts," and the bearing of this doctrine upon a possible re-analysis of the cosmological argument should be examined. Although it seems to me that the objections I have brought against any doctrine of "epistemological objects" are also objections to this kind of philosophizing, I shall try to deal with the new arguments in an independent way. All the same I should like to record my conviction that there are no "speech-things" or "things-as-in-speech" (except words) just as there are no "thought-things" or "things-as-in-thought" (except cogitations).

According to Mr. Wisdom[1] "Incomplete facts are those which are expressed by incomplete sentences". This statement, I think, should make us stop and reflect. If by a "fact" we mean, as we usually do mean, "that which makes an existential assertion true", it is difficult to see how an incomplete sentence could express a "fact" in any relevant sense; for it could not express anything that makes any existential assertion true. Consider, for instance, such an incomplete assertion as "Tom is balder". That is the beginning of an assertion that, as some novelists would say, has trailed off into silence. There is no "fact" expressed at all until we are told that "Tom is balder than James", or that "Tom is balder than he used to be" or something of the kind.

Extending his argument Mr. Wisdom says that there are "incomplete facts" when anything particular is designated generally or indefinitely. General statements, he allows, may

[1] *Mind and Matter*, p. 27.

The Cosmological Argument

be complete when they refer only to what is general, as when we say that orange is between red and yellow. A general designation of what is particular, however, is always incomplete, and an indefinite designation is always incomplete. According to Mr. Wisdom we have an incomplete fact when we say, for instance, that someone is happy. It is not the indefinite "someone" who can be happy but definite human beings such as Wang, or Paul Jones or Robert Louis Stevenson.

It appears to me that the statement "Someone is happy" is not an incomplete assertion; for it may be true. Indeed if Wang or Stevenson be happy, it must be true, by what is technically known as a weakened conclusion by immediate inference, that "someone is happy". At this point in his argument Mr. Wisdom himself appears to be dubious. We are, he says, "tempted to say"[1] that such a statement is an incomplete fact although "it would be better to say that it incompletely expresses an ordinary complete fact." I submit that the temptation should be resisted, and the "better" way of speaking alone employed. It is quite fallacious to argue that because we may refer to things indefinitely, therefore there are indefinite things. It is also fallacious to argue that if we refer generally to something that is particular we thereby convert that particular thing into a generality. It is quite usual to believe that the nature of any existent thing is inexhaustibly rich and the belief may very well be true. If so, any reference to the thing by means of a selection of its properties would be incomplete; but it would not generate an incomplete thing.

I submit, therefore, that there are no incomplete facts, and consequently that there is no anchorage for a re-analysed cosmological argument in precisely these waters. The topic, however, is connected with the contentions of the finitists in metaphysics, and in that respect may require very close attention.

I have said that there are no incomplete facts, no incomplete things, no incomplete realities. It does not follow, however,

[1] *Mind and Matter*, p. 28.

Theism and Cosmology

that we may not frequently use expressions that profess to designate facts, things or realities, but turn out upon interrogation to be incomplete or insufficient expressions. Such expressions would pretend to a virtue that they do not possess, and our question now is whether "nature", "the world" or similar expressions for the existential premiss of the cosmological argument may not be of this pretentious kind. If so, what follows?

Grammatically we may speak of the world as we speak of a table or of a chair; but are we entitled to speak of the *world* in that sense? Do we know that there is such a thing? Again, if we are in sympathy with those philosophers who deny that there is any one existent entity correctly to be called a table, and affirm that all our references to existence should be to something much simpler than a table—let us say, to "this noise"—have we any right to say that "the world exists" in the sense in which we have the right to say that "this noise exists", or in any other defensible sense?

According to the doctrine of finitism there are no genuine existents (or, indeed, objects of thought) except such as can be discerned as individual entities, and ticked off by enumeration. This doctrine, in the extreme form in which I have described it, would seem to be far too sweeping to be admissible. It would involve the consequence, for instance, that we were talking about nothing intelligible if we spoke of all the centenarians at present alive on the earth, supposing, that is, that birth certificates, or other such convincing evidence, were not always available. In that case (it seems plain) we know that there is such an empirical group although we cannot enumerate its members with complete precision. On the other hand it can scarcely be said that if we tried to talk about "all the integers there are" we would necessarily be talking sense. We may be able to talk significantly about *any* integer without being able to talk significantly about the totality of the integers, that is, about all the integers there are.

Hence it may be maintained that while it is reasonable to

The Cosmological Argument

talk about *any* natural event as well as about *this* natural event, it need not be reasonable to talk about the "world" if by the world we mean the totality of things, all the things there are, or even if the "world" was what Mr. Wittgenstein said it was in the second aphorism (1. 1) of his *Tractatus Logico-philosophicus*, "die Gesamtheit der Tatsachen, nicht der Dinge"—"the totality of facts not of things".

We are accustomed, it is true, to speak of the "universe", the "world", or "nature" in the belief that there is just one natural universe, with a unitary ground-plan. Monists and pluralists dispute about the question whether the universe is one thing or a group, whether its constituents may be radically disparate or are varieties of the same sort of stuff. But now we are asking a much more fundamental question. The dispute now is whether we are entitled to speak about "the universe" at all. According to the extreme doctrine of finitism we are only entitled to speak about this, that, or the other event, fact or thing, not about the totality of any or of all of these. According to the more moderate opinion which (I have suggested) would have to be admitted as an indispensable supplement to extreme finitism, we are further entitled to speak about *any* event, or fact, or thing. But that is not a proof that we can intelligibly speak of "the All", the "universe" or the "world".

If I were challenged regarding my meaning when I speak of the physical universe, I should say, primarily and very roughly, that I mean to assert that everything that has been, directly or indirectly, sense-evidenced belongs to a single spatio-temporal order that is patterned in a distinctive way, usually described as causal. This single order, I should further say, may have had an infinite past and may have an infinite future. Nothing in the statement denies its infinitude, although neither its finitude nor its infinitude need be asserted.

In this general sense I have spoken and, like other people, will continue to speak about, "the world". Such a form of statement, I think, might perhaps be translated without any loss into the language of modified finitism, on the general lines described above. It could not be translated into the language of extreme finitism, but I have already tried to show that extreme finitism fails to express what can and must be intelligibly expressed.

Therefore, for all I can see to the contrary, it may be true that the existential premiss of the cosmological argument speaks about the "world" in a sense that could not be readily defended against a hostile critic. We may therefore enquire whether, if this were true, the cosmological argument could adjust itself persuasively to such a situation.

The question would now be whether the "world", translated into the language of a modified finitism, yielded an inference, by complementary relation, that either demonstrated or supported a theistic conclusion.

The answer, I think, is plainly in the negative, unless "God" be a name either for a being sense-evidenced or evidenced in a way similar to sense-evidence, or definable as "anything that is so and so". In most of the senses in which philosophers and theologians have spoken of God, the arguments that challenge our right to speak about "the world" also challenge our right to speak about its God. If God be a name either for the Whole or for a hyper-cosmic whole, the modified finitism that we are now considering denies that we can intelligibly assert the existence of any such whole, secular, deiform or of any other kind. From the existential premiss there is no inference by complementary relation.

The principal results of the minute enquiries into wide, and sometimes into thin generalities that I have attempted in this lecture, are I think, two. The first is that if the existential premiss of the cosmological argument is expressed in a highly general, neutral way, the inference from it is either indefinite or distressingly emaciated. Unless the argument proceeds from distinctive, if still very general, characteristics of worldliness

The Cosmological Argument

it cannot hope to reach a distinctive over-worldliness. Yet the argument as I have said, and repeated, professes to argue generally from the debility of the world at its proudest. The second principal result is not dissimilar. It is that the inference from complementary relation employed by the cosmological argument is apt to rely upon correlatives that do not exhaust the logical field but, far too frequently, are little better than a vague analogy.

As I said at the beginning of this lecture, the cosmological argument is the most general of the traditional theistic proofs *a posteriori*. Therefore it seemed to be a suitable starting-point; for the field of theistic argument *a posteriori* is very wide, and very relevant indeed to natural theology. The mere plan of the field, in its abstract form, may be flat and dull. But it is well to have a plan, and a large proportion of the arguments that we shall discuss in the next few lectures may be regarded as supplementary to the bare generalities of the cosmological argument. Even if the whole of this lecture had been employed in weeding an abstract field, it need not have been wasted; for weeding is an indispensable if soulless and stooping part of metaphysical gardening.

It may be asked whether the attempt *was* indispensable, whether these weeds are anything more than the ghostly crop of bygone metaphysics. To that (I submit) the answer is that if the objection be to metaphysics it is an unphilosophical objection. What else should a philosopher want? If the objection be to *old* metaphysics, it is also unsound; for old metaphysics, like mellow wine, may be better than the new. If, however, the objection be that *bygone* metaphysics, like wine that has decayed, should not appear at a modern table, it would be a very good objection provided that the disparagement of such metaphysics or of such philosophical theology was due, not to neglect, but to assured knowledge of the decay. What we find in fact (I suggest) is something quite different. Metaphysics of this type has become unfashionable. It is known to have aroused dubieties in its own circles, but

Theism and Cosmology

it is only presumed to have been refuted; and the same type of argument is served up again without any recognition that it is a twice-cooked dish. That is a disturbing situation, and I can think of no better remedy than a review of old-fangled contentions, not necessarily in their oldest forms.

IV

Creation

THE late Dr. Farnell remarked in his book *The Attributes of God* that "only in a very few of the more advanced religions has the idea of cosmic creativeness been attached as a primary function or as an essential attribute to the High God".[1] The advanced religions in question, he said, were Judaism, Mahommedanism and Zarathustraism. Creation myths, he says, were present in (although not essential to) other religions, as in the story of Marduk in Babylonian mythology. In another part of his book he gave an illustration from Peru. "The creator Pachacamac made all things by his word; let earth and heaven be";[2] but Dr. Farnell did not discuss the prominence of this idea in ancient Peru. From Egypt he collected the astonishing tale that the God Neb-er-tcher "desirous of creating the universe, first uttered his own name as a "word of power" and then evolved himself and all the world".[3]

Dr. Farnell also pointed out that in the monotheisms of Judah, Arabia and Persia, the creator was not a nature-god, but a spiritual and moral entity. So was Marduk, although Marduk was not a very pleasant being.

That may suggest the inference that the idea of a creative God has its roots in moral and humanistic rather than in cosmological grounds. If so, Egypt, as usual, provided an interesting commentary. Re, the sun-god, was said to have "made heaven and earth at man's desire"[4] and Re (who was the Highest God) was also regarded as the fountain of justice.[5] For some of the Pharaohs cosmology was primary and also

[1] p. 115.   [2] p. 232.   [3] p. 239.   [4] p. 115.   [5] p. 141.

Theism and Cosmology

moral, a striking instance being the scientific sun-worship of Amenhotep IV.

This piece of history need not have much importance for the philosophy of theism, but, in a general way, it may be suggestive. If many of the advanced religions regard creation as an essential attribute of deity although naïve theology does not, and if the natural man (as we call him) is a worshipper but not of his Maker, it might still be reasonable to believe that the idea of creation is vital to theism although it is a sophisticated and not a crude notion. Let us consider the hypothetical question why an advanced religion should be driven or inclined towards the sophisticated idea of world-creation.

In the first place the reason might be that a philosophical (which must be a sophisticated) theology gradually came to perceive that it could not afford to be less than cosmological, and so evolved one of the boldest of all cosmological theories. The world is far too big to be ignored, and God would not be a God unless he was stronger than the world as well as higher than man. If it took many millennia for mankind to perceive this truth in clear, sharp outline, the fact need surprise no one. The profounder simplicities may be hidden for centuries of human generations.

Again, if it be true that the motive to theism in most of the advanced religions is humanistic rather than cosmological, and that these religions tend to regard deity rather as a magnified non-natural emperor, war-lord, shepherd or comforter than as the maker of the universe, it becomes all the more fascinating to observe how the need for combining these super-humanistic attributes with an over-worldly cosmological power gradually overmastered some men's minds. That, indeed, is the crux of the problem of "science and religion" in these modern times. If science and religion were simply different, each might go its several way. Theology need not instruct the sciences. Even if she be the Queen of them all, she need not be more than a scrupulously constitutional monarch;

Creation

and philosophical theology need not be concerned with the scientific accuracy of the statements in any sacred book. Theology might still have the gracious duty of delivering mankind from the tyranny of myopic this-worldliness and of opening the gates to the mansions of hope. There would be a wholly different situation if science and theology *must* be joined together. In that case we could not contemplate an inevitable union between partners who have no common language, no common memories, no common traditions, no common ideals. Yet such would be theology's destiny if it neglected cosmology. That is the reason why the bolder theologians still maintain that science should not persist to the end in its attitude of *Nolo episcopari*.

I am not suggesting that the idea of creation provides the only possible basis of union. The denial of all creation, or more narrowly of world-creation, need not be the ruin of this partnership, and estrangement would be possible even if the idea of world-creation were admitted to have very strong cosmological grounds. I do say, however, that theism, on its cosmological side, would be very greatly strengthened if it could be shown that the world required a creator or creative society. That statement, in its turn, requires a close examination of the theory of cosmic creation not as a suggestive parable but as a piece of sober philosophical argument.

The importance of such an enquiry is, of course, that if the creation of everything except God himself were held to be an inevitable verity, and if God's creative power were combined, as it is usually said to be combined, with an equal and similar power of annihilating everything except himself, then the kingdom of heaven could not suffer violence from any *external* source. There could scarcely be a theological proposition of profounder import.

I have put the point at its highest elevation, describing it in terms of the creation or possible annihilation of *everything* except God. There could also, of course, be theories of a limited and partial creativeness. Even poets may be creative,

Theism and Cosmology

makers in their own tiny way. For the moment I am not speaking about that but about the creation of the world. Such creation need not be entirely devoid of restrictions. It would, for instance, involve the consequence that not even God could annihilate the past. If he made the world and unmade it again it would still be true that he *had* made it, that he had brought the temporary fabric into existence although he later thrust it out of existence. That, however, is part of the meaning of production rather than a limitation of it; and the denial of all *external* limitations to God's power remains most profoundly important. The non-natural creation and non-natural power of annihilating the world carries the implication that if there be any weakness in this matter the defect lies in God himself and not in any exterior obstacles that might conceivably hamper him.

This implication, it is true, may have its own embarrassments. It is God's presumed omnipotence that is the scorpion's sting in the venerable theological problem of evil, and the removal of all external limitations to God's power comes very near to being an assertion of his omnipotence. "I am the Lord", it is written in Isaiah (xlv, 7–8), "and there is none else. I form the light and create darkness; I make peace and create evil; I the Lord do all these things." That is a terrible text. It may suggest that the voice of nature proclaims a God that the voice of human nature deplores. Nevertheless it is at least intelligible that a theism that may be terrible is better than no theism at all; and our problem at present is not about what is terrible, but about what is true.

Let us proceed, then, to examine the doctrine of the creation of the world, its meaning, its reasonableness, and its relation to other theories allied and hostile.

It is a matter of common knowledge, I believe, that the accounts of the creation of man given in the first and in the second chapters of the Book of Genesis betray an inconsistency of ideas. Even unlearned persons may have heard that the second chapter incorporates an older story"J. E.", while

Creation

the first is the later *Priesterkodex*. However that may be, it is plain that there is a discrepancy regarding the *modus operandi* of the originating events. In the first chapter the operative agency was a mere volitional *fiat*. "And God said, Let us make man in our image, after our likeness." In the second chapter we read, "And the Lord God formed man of the dust of the ground, and breathed into his nostrils the breath of life, and man became a living soul".

It is not my purpose to examine the logic of the Hexateuch, and I have intermeddled with the beauty and restrained simplicity of the story in the early chapters of Genesis for the sole purpose of acclimatizing our conceptions. Plainly, however, there is a world of difference between the conception of God the Potter and the conception of God the Creator. The potter may be a limited creator in the sense in which a great artist may be a limited creator; but he operates upon pre-existent material and is limited by the capacities of such material. Without the Potter we might perhaps be able to say "No world without God", a formula sometimes used to describe the essentials of the creation theory. For without the Potter there might only be shapeless, unordered clay and not an ordered world. According to the creation theory, however, we ought to be able to make a much stronger statement, the sort of statement indeed that, as we have seen, is the grand theistic motive to credence in the creation theory. We ought to be able to say that there is nothing *except* God and the mere effluence of his power. The theories of creation and certain forms of the theory of emanation would enable us to make this grand assertion. The theory of God the Potter would not.

Indeed, so far as I can see, there are three and only three theories according to which the immensely important proposition that there is nothing except God can be intelligibly maintained. One of these is pantheism in its various forms, and although it is a tautology that, according to pantheistic theories, there can be nothing except God, it is legitimate to point out that pantheism need not, and as an expression of ultimate metaphysics cannot, include any theory of origins. It is the *prius* of all origins. If a theory of the origin of the world is included, then, if there be nothing except God, we are limited either to a form of the emanation theory, or to a theory of pure creation, creation *ex nihilo*. The emanation theory need not include a theory of temporal origins, for the so-called emanations may be regarded as eternal strata in God's eternal being, but it is consistent with a theory of temporal origins, namely that the world was spun from the divine substance. The creation of the world *ex nihilo* (combined with the possibility of its reduction *in nihilum* by divine *fiat*) is the third possibility, perhaps less strictly all-divine than the others, but still substantially a theory of omni-divinity. To speak summarily: On the hypothesis of omni-divinity either the world *is* God, or God made it out of his own substance, or he made it, but not out of anything (as Bowman has said, the most accurate way of stating the doctrine of creation *ex nihilo*[1]). On the two latter hypotheses there is a sense in which either an emanation or a created product is a distinct existence when released or created, but that may be of small account if it be expressly maintained that when the mystery is accomplished, the temporary world-emanation will be absorbed once again into the divine being, or alternatively that the divine work of creation will be divinely undone.

I cannot see that the creation theory has any important theological advantages over the emanation theory. At the most it emphasizes God's transcendence at the expense of his immanence. I was therefore surprised to find in that very liberal-minded document *Doctrine in the Church of England* the following exceptionally strong statement: "Any such view as that the finite universe proceeds by emanation from the Divine nature, as opposed to the view that it originates in the creative activity of the Divine Will, is non-Christian."[2]

[1] *Studies in the Philosophy of Religion*, II, 427.   [2] p. 45.

Creation

I suppose it might be argued that there is an important difference between saying, as emanationists might say, that the world is an expression or utterance of deity, and insisting that it is a voluntary utterance. It would seem to me, however, to be very odd that Christianity which in its ethics lays so much stress upon meekness, purity of heart and other qualities of character that cannot be summoned at will, should lay so much stress upon divine volition as opposed to other forms of divine self-expression; and this point of difference would seem to be very anthropomorphic indeed. In Christendom the important Monothelite controversy was about a single divine energy—μία θεανδρικὴ ἐνέργεια. Mahommedans, again, put the emphasis upon God's holy unitary power and not upon his "volition" in a narrower sense.

Mr. Webb, regarding all such theories as a piece of profound symbolism but still only as symbolism, says that creation better expresses the difference between God and man than either generation or emanation. Of the two latter he further says that the notion of generation is preferable to the notion of emanation "in that the latter suggests a process more wholly unconscious and involuntary than the former".[1] I cannot see that there is much health in these contentions, even regarded from the point of view of symbolism. The emanation theory, it is true, may be described in terms which do not imply consciousness. A favourite neo-Platonic illustration used by Origen among others[2] was drawn from a mistaken analogy from physics, namely that the sun imparts its light and heat without diminution of its own substance; but it would be unreasonable to restrict the emanation theory to such analogies, or to burden it with the weight of an abandoned physics. Indeed many emanation theories, and all of the Logos kind, are epistemological rather than physical. Again, while the theory that the world was begotten, or that it arose from generation, may have some slight suggestion of voluntary action, it would surely be appropriate to regard such theories

[1] *God and Personality*, p. 162. [2] *De Princ.*, I, 2, 4.

primarily as *biological* theories of emanation. Moreover their resemblance to the monstrous imaginings of many religious myths, as in the figure of a more than gigantic egg, or of a huge spider spinning the world from its entrails, is very obvious, and need not supply greater evidence of intelligent origination than the growth of a mushroom. As it seems to me, there is little to commend such arguments.

According to Dr. Tennant the emanation theory is defectuous in comparison with the creation theory because it loads the ultimately inevitable mystery of world-origination with unplausible suggestions concerning its *modus operandi*. "The notion of creation", he says, "is not derivable from experience, and analogies valid within experience cannot reach beyond its powers."[1] That is an important matter. If the emanation theory cheated us with suggestions of verisimilitude, and if the creation theory avoided all such shams, the advantage would lie with the latter. It is hard to see, however, why the theory of a creative volitional *fiat* should be regarded as exempt from such suggestions. Indeed Dr. Tennant, on the very next page, tells us that volitional creation "minimizes the inexplicability of things" because it helps us to explain "an imposing array of considerations that bespeak design". What is this imposing array except an array of analogies within experience?

Again there is something (although, I daresay, not very much) to be said on the other side. Both the creation and the emanation theories are equally vulnerable to the objection that it is unintelligible how a changing world could arise from a changeless deity, or again that there is no intelligible reason why the world should begin at one time rather than another, there being nothing outside God to make any moment more propitious than any other. On the other hand, if God and his universe are co-eternal, we might speak with Victorinus of *semper generans generatio*, and might think it was aesthetically more fitting that the world should be in a manner divine, than that it should be, like a work of art, quite distinct from its

[1] *Philosophical Theology*, II, 125.

Creation

maker. Again, if the objection be that God, being perfect in himself, cannot have any sufficient reason for originating what must be inferior to himself, it would apply equally to an inferior emanation and to an inferior created product. Nevertheless there might seem to be less difficulty in the conception of a region of divinity eternally consubstantial with other regions of divinity (and an "impiety" only if it were falsely supposed to stand alone), than in the conception of an inferior creaturely universe existing even temporarily alongside of God.

The term "proceeding" was also used. Indeed, the chief official difference in doctrine between Eastern and Western Christendom is whether the Holy Spirit "proceeds" from the Father or from the Father *and* the Son—the famous *filioque* clause. In the Athanasian creed we are told: "The Father is made of none: not created nor begotten. The Son is of the Father alone; not made nor created but begotten. The Holy Ghost is of the Father and the Son; not made, nor created, nor begotten, but proceeding." Here the word "proceeding" I should say is rather vague, some of the others rather too definite in any simple sense.

It is hard not to sympathize with Gregory of Nazianzus[1] when he said: "Do you ask what is meant by the procession of the Spirit? Tell me what you mean by the Father being ingenerate, and I will give you the physiology of the Son's generation and the Spirit's procession. Who are we that we should handle such matters as these?" Unfortunately, however, these very abstract questions of very abstract "physiology" were just Gregory's problem. I can see only three clear distinctions; (1) that the world is made, but not out of anything (i.e. creation), (2) that the world is made out of something (i.e. that undivine material, perhaps chaos, perhaps "space" is ordered by God), (3) that the world is some sort of output of divine substance, (i.e. a literal "proceeding" or emanation). With that remark I propose to leave what seems to

[1] Quoted by Headlam, *Christian Theology*, p. 419.

Theism and Cosmology

me to be a sterile promontory, and to return generally to the hypothesis of creation, the usual theory in our tradition.

Let us begin by considering whether there are any instances of creative process *within* the world, and, if so, whether such this-worldly creation is of any service towards diminishing the mystery of the creation *of* the world. In asking this question I am not forgetful of the warning that Hume, Dr. Tennant and others persistently give. It is illegitimate uncritically to apply intra-mundane analogies to supra-mundane cosmology. The emphasis here, however, should be on the adverb "uncritically." It is equally uncritical to assume that no conception that applies *within* the world may not also apply *to* it, that we must either have an entirely over-worldly faculty for cosmology, or else be intellectually as well as verbally dumb regarding that theme. That, to say no more, should not be the attitude of a philosophical theology. On the contrary, just as many mathematicians attempt to discover a conception of number that applies both to finite and to infinite numbers, so philosophical theologians should be put upon their mettle when they are reminded (or remind themselves) that analogies *within* the world need not apply *to* it. They may ask, for instance, whether there may not be world-determinants that are not "natural causes" in the ordinary sense, but nevertheless have a common essence with such causes. The answer may be "Yes" or "No", but the thing should at least be enquired into. Again, although the *sensus eminentior* of theology may sometimes be more convenient than convincing, it is not, on that account alone, a phrase that is quite forlorn and wholly desperate. In short it is not inconceivable that creative process within the world, if there be such process, may provide a cosmological clue.

One of the most frequent uses of the term "creative" in ordinary discourse concerns the imagination. Thus "creative" imagery is distinguished from reproductive. Hobbes called the latter "decaying sense"[1]—not inaptly, since the freshness

[1] e.g. *Elements of Law*, chap. iii.

Creation

at least has gone, and waste, if not destruction, can be said metaphorically to have occurred. Hobbes's language, however, is very questionable indeed, if its meaning be that earlier first-hand experience literally persists in the so-called "reproductive" imagination; and even the most slavishly repetitive imagination varies to scale and in other ways.

Illustrations of "creative" imagination are naturally sought in the domain of art, and especially of fine art.

If we consider the products of the finer arts, such as a painting or a statue, or a piece of tapestry, or a poem, it is plain that these, regarded as material or as matter of fact, are not created *ex nihilo*,[1] since they are rearrangements of previously existing material, pigments, marble, threads or words. It may however be maintained that such physical materials are essentially the cues to aesthetic experience, the occasions and not the substance of an experience that *is* "creative". This might be said of the beholder as well as of the "creative" artist, although it is usually applied to the latter only. The beholder, even if he is a "Raphael without hands" or a mute unpractised Milton, may, for all his deficiencies as a skilled executant, be a "creator" in his enjoyment.

To this it is sometimes replied that neither the artist's nor the beholder's imagination is exempt from the common lot

---

[1] Mr. Collingwood (*The Principles of Art*, pp. 129 ff.), choosing his examples from "non-technical" creation—e.g. "creating a disturbance or a demand or a political system"—concludes that "creative" in everyday English may be described negatively as action (possibly responsible and voluntary) which need not be directed towards an ulterior end, need not follow a preconceived plan and "is certainly not transforming anything that can properly be called a raw material". In positive terms, according to this author "creation" is involved in "total imaginative experience". This he describes without prolixity in about 70,000 words, and I shall not try to condense the exposition. "Total imaginative experience," according to Mr. Collingwood, may originate "real" things, but is not *ex nihilo*, and he repudiates all theological inferences. In that I think he is wise. If the world were God's language or utterance it might be God's "total imaginative experience" in Mr. Collingwood's sense, but I should despair of being able to find a place for humanity in such a world. I take it that Mr. Collingwood agrees, and consequently is not a theist on these lines.

Theism and Cosmology

of all imaging. All imagery, the objector avers, is but rearrangement of old material. Those who are born blind cannot dream in colours. Therefore, visual imagery borrows its constituents from sensation and does not create them. Indeed it has been plausibly suggested (although there can be no definitive proof) that the same is true of all the finer nuances of colour as well as of the grosser colour differences. Similarly of the auditory images in the mind's ear, the taste-images in the mind's palate, and so forth.

This argument strongly suggests that if "imagination" be restricted to actual imagery then the mind's "creative" power in the way of the creation of images is very limited. True, it may be replied that there is as great a difference between the image-constituents and the aesthetic product as there is between tubes of paint and El Greco's "Burial of the Count Orgaz". Such a reply, however, need not yield strong indications of literal creativeness. The new fact is not made out of nothing. It is not very convincing to suggest that it is made *in part* out of nothing; and any collection may have properties that do not belong to its individual constituents. Even if such properties dazzle the mind with a new and sudden glory, or rouse it to aesthetic imputation of the highest order, these facts, in cold logic, are not productions out of nothing.

Still, it may reasonably be contended that imagination is wider than imaging, even if it always contains some imagery, and so that the formation of images and of image-patterns is but a small part of imagination in its wider, more genial meaning. There was an easy play of images in Shakespeare's tavern talk about "this bed-presser, this horse-back-breaker, this huge hill of flesh," but all such images were so much less than his sweet Jack Falstaff, that we need not regard them as much more than lively embroidery. Let it be granted that Shakespeare drew upon his memories of soused gurnets, and ancients, and revolted tapsters and ostlers trade-fallen. The provenance of such images in his plays may have as little bearing upon his creative genius as the fact that the words

Creation

he used had, almost always, been used by others in different contexts.

Indeed it is very hard to deny that people do mean something when they speak about creative work. If they call themselves creators they mean that they are not intentionally imitating or *re*producing. They believe themselves to be the authors of something new, with a novelty they believe to be wholly misdescribed if it is said to be a combination of old elements in a connection hitherto unnoticed. It seems to them to be vitally fresh in its imprescriptible integrity. They are (or they often seem to themselves to be) spontaneous agents in such imaginative production, and although they may not think that they are creating out of nothing, they are confident that their imaginative products are quite distinct from any conscious antecedents in their own minds. There is a marked contrast, in their very vivid and very personal experience, between work of this imaginative kind, and the rest of their work. Therefore "creation", on their part, is significantly contrasted with much else that they do, and the word "creation", regarded as a contrast-term, may be more expressive than most.

On the other hand, a contrast-term that has *some* empirical meaning need not, on that account alone, literally have the meaning that it appears to have, and may be quite unsuited for technical philosophical purposes. Artists describing their own experience need not be taken to mean either that *they* "create" or that they "create *ex nihilo*". They might say, as many of them have said, that they did not know how their imaginative ideas arose, and that they did not even know that *they* had produced them. All they could say with certainty might be that these ideas somehow turned up in their minds, and many of them have been convinced that such ideas were the stirring of a breath that came from outside, that they were the privileged if humble instruments of an afflatus that was not their own, perhaps God's, perhaps, as for Mr. Yeats, the surging overflow of an immemorial racial treasure.

Accordingly it would seem to be impossible to extract very much from what is called creative imagination in human artistry. There is too much doubt about the sense in which it is creative. If that comment has to be made on the great creative artists it would hold, *a fortiori*, of the sense in which all of us, in nearly all our work, are largely compact of imagination. What is doubtful in the more eminent instances is still more doubtful in the shier ones.

Continuing this attempt to discover creativeness in some special sense within the world, we may remark that procreation, life and growth are sometimes regarded as creative in a quite peculiar way, and indeed that psychological creativeness is sometimes said to be only a form of vitality. According to the theory of vitalism, freely interpreted, living things are centres of spontaneity, limited indeed as regards the range of the novelties they can achieve, but within these limits ahead of their environment. They are not to be regarded as a sort of clearing-house or telephone exchange in which everything that comes in by one entrance goes out by another. Life is always a forward movement; and that may be another way of saying that it is always creative. If not, why not?

Many problems cluster in these assertions, as we shall see in due course. For the time being we are concerned with the assertion that life is creative in a sense in which lifeless existence is not creative. That assertion appears to be vague. No living thing is created *ex nihilo*. (Probably it has at least one parent.) Its transactions with its environment are transformations of energy in which the ingoings and the outgoings balance one another with very marked experimental precision. It has to live the life of its kind, a man a man's life and a dog a dog's life, despite a certain pliability in the evolution of species.

That is what an external observer would have to say. If it be replied that "spontaneity", the sense of novelty and of authorship in living, depends largely upon a metaphysics of inside information, we have to ask whether the supposed inside information is necessarily veridical; and it does not

Creation

require very penetrating eyes to see that our inside information about our own spontaneity may be specious only. No one is greatly surprised, and very few are impelled along the road to general scepticism, by discovering that what appear to be our most vagrant and motiveless actions show marked unsuspected regularities or other evidence of secret causes. Most of us, indeed, come to see that if our so-called sense of spontaneity were illusory, our lives would not look very different, even to ourselves, from what they seemed to be before the illusion was detected. Our theories about them might be appreciably altered; but that is another story. There is nothing approaching a proof that "spontaneity" correctly describes the absence, as opposed to our mere ignorance, of relevant and of sufficient antecedents.

Therefore I think we should infer that so-called biological creativeness is not very encouraging to a theist who is searching for something dimly analogous to the sort of creation that his cosmology needs. I also think that the empirical contrast between dead and living matter is of little avail. It is generally agreed that there is no such thing as a peculiar vital substance that acts as a dynamo and runs a machine whose constituents may all be found among dead things. Therefore the contrast between dead and living matter has to be developed along other lines, and I think upon lines of a much broader gauge. It may even be suggested that life, so far from being simply a temporary characteristic of a few scattered beings on the earth's surface, should be regarded as a general tendency within the universe peculiarly manifest in animals and in plants. If so, nature, although not herself a living thing, would contain the rudiments of a spreading vitality. Nature may also, no doubt, contain the rudiments of a spreading mortification. That, if it were true, would not contain any contradiction. If plants and animals suggest the presence of a life-like cosmic tendency, granite and steel may suggest the presence of a death-like cosmic tendency. And why not?

Theism and Cosmology

This leads to a third empirical consideration regarding creativeness. "Matter" we are told is now regarded less massively, and is even volatilized. Life, in many quarters, is regarded less soulfully. Therefore the old-fashioned dualism between a set of billiard-balls and a mob of sprites is out of date. Indeed, nowadays, it is common to credit the electrons with very spritely properties indeed, and to think of them as individually "spontaneous" although collectively "determined".

Creative electricity, in short, may be supposed to be at least as plausible a theory as creative biological evolution. Creativeness, it is true, must be more than a mere negative like indeterminism. Nevertheless the supposed indeterminism of the individual electrons implies that their behaviour is not *pre*-determined in detail. All the electrons are like the celebrated amphisbaenic snake:

> Until it starts you never know
> In what direction it will go.

If so, the behaviour of the individual electrons must be independent, in part at least, of all that went before in the world. Such behaviour, we may be told, is either self-caused or uncaused. In the former case we should have a type of action that may look very like creation. In the latter case, on the other hand, we should have to decline to present any explanation whatsoever.

That is all I propose to say here about empirical creativeness in some special sense. It does not seem to me to help a theist who believes in the creation of the world. The discussion, however, has been heading towards a much weightier matter of metaphysics, namely towards the view that creativeness instead of being a special characteristic of a few peculiar entities within the world is a universal world-characteristic, and also a universal characteristic of any over-world there may be. For the rest of this lecture I shall be concerned with

Creation

this wider problem and with its relation to the sort of creation that some theists demand for their cosmology.

The type of theory now to be examined is at least as old as Heracleitus, and has seldom been inarticulate since his day. One remembers the lines of Lucretius[1]:

> Mutat enim mundi naturam totius aetas
> Ex alioque alius status excipere omnia debet,
> Nec manet ulla sui similis res; omnia migrant,
> Omnia commutet natura et vertere cogit.

The present age, however, following Darwin and impressed by M. Bergson, may have been more profoundly impressed by the time-series than most other ages, and may have had less avidity for the transmutation of the time series. In our own English language, Mr. Whitehead has made this problem peculiarly his own, and I shall be thinking of his views in *Process and Reality* in most of what I have to say. I shall not, however, attempt to give either an exposition or a criticism of his philosophy.

What is primary in this conception, as I apprehend, is that "creativeness" or "creative advance" is part of the very meaning of process or becoming, and that everything that exists must be something that *goes*. To occur is to come into existence, to *be*come, and anything that comes into existence is about to go on. In all process, transition or becoming there is always passing *away*, but there is always passing *on*, and, one might say, *pressing* on. The latter aspect is the inherent creativeness of a reality that proceeds.

So far, I believe, Mr. Whitehead would agree. I have not the same confidence that he would agree with what I am about to say; but I also am passing *on*. The present I suggest is always parturient, and nothing is parturient except the present. There is nothing *doing* except in the present. It succeeds the past, but the past, in virtue of having passed away, is impotent.

[1] *De Rerum Natura*, V, 828 ff.

Theism and Cosmology

There are certain senses in which these statements would be utterly tame and completely innocuous. If the term "exists" or the phrase "is actual" is in the present tense, grammar without metaphysics suffices to show that the past should be indicated by the past tense and the future by the future tense. The view I am suggesting, however, has more than this trivial implication. It asserts that tense is ineluctable in all determinate statements about determinate existence. The past, at any moment, is determinate but impotent. It cannot be undone, but it cannot do anything and it remains impotent for all later time. The future at any moment is indeterminate before "it" occurs. When it occurs it will become determinate, and it will remain determinate ever after; but before it occurs it is nothing determinate; and that, ultimately, is the reason why no man can foresee it in the literal sense of foreseeing. Before it occurs it can do nothing; for there is no such thing as an indeterminate agent. As opposed to past and future, the present, at any moment, has quite peculiar features. It is determinate and potent, thus differing from the past, which is determinate and impotent, and from the future, which is both indeterminate and impotent. These statements are not tame and they are not innocuous. On the contrary, they imply profound metaphysical and theological consequences.

Plainly they bristle with difficulties. One may ask whether they repudiate the modern doctrine of local times, and so are inconsistent with the theory of relativity. One may ask whether the present could be intelligible on such a theory. Would not the theory interpret it as a dimensionless point? How could all agency be confined to such a point and how could the dimensionless point combine the properties of passing *away* and also of pressing *on*? One may ask whether the theory does not imply that everything has the incompatible properties of pastness, presentness and futurity (in the tenseless sense of "has") since everything now future will become present and later will become past.

These are only a few of the difficulties, and I shall postpone

Creation

consideration of most of them until I come to treat of eternity and its relation to time in the next lecture. In the present lecture I am content to allow the discussion to be in part provisional.

In Christian theology the assertion is that the world is the work of a divine producer who has nothing outside him to work upon and (if the emanation theory be repudiated) does not release any part of his own substance when the world comes into being. The world, therefore, is produced or originated. It has a source and is not itself the home of all its origins. Because it has a source it does not simply pop into being and perhaps pop out again. It is made by a maker.

Such a theory is very different indeed from the view I have attempted to sketch, namely the absoluteness of becoming as such in all existence, even God's.

In one respect there is partial agreement (although only partial) between Christian metaphysics and the theory I have attempted to sketch. Both agree that "creative" process always proceeds from a source. The term does not imply utter novelty in what is said to be created. It does not imply complete discontinuity, or mere popping into being. On the other hand, while Christian metaphysics says that God is the source of the world, the theory I have sketched asserts that what is actual at any moment is, as such and altogether, the source of what succeeds that moment.

I shall first examine what both theories repudiate, namely the view that anything simply pops into existence.

This phrase, which is colloquial but not intentionally facetious, obviously contains a metaphor. The popping of corks or of paper bags has a cause and does not come from nowhere. Such noises are sharp, and sudden and unexpected; but that is all. I am speaking therefore of metaphysical or pure popping, meaning to say that there is a definite and complete absence of continuity. In that sense the theory of intrinsic creativeness that I have sketched denies pure popping. It emphatically asserts the continuity of what passes *on* with what passes

Theism and Cosmology

away, or, as is sometimes said, the universal connection between creativeness and inheritance.

It may be complained that any theory of pure popping is not an explanation but the despairing and mean-spirited renunciation of any explanation. That may be true. It cannot, however, be dismissed as defeatism and poppycock unless there is some way of demonstrating the necessity of a cause for every occurrence, and that, in most modern theories, is held to be beyond human wit.

It is therefore important to distinguish between creativeness in the sense I have explained, and an assertion of mere popping. This is often forgotten. Thus in certain philosophies the assertion seems to be that "creativeness" is simply "novelty" defiant of all inheritance. "We have denied one hundred per cent inheritance", I read in a recent book.[1] The meaning would seem to be that some percentage of process is uninherited creation, and that this percentage is pure popping. If so, this would seem to me to be one of the oddest of all strange metaphysical prescriptions, and to have little or no relation to "novelty" in the sense that is wanted. There might be any number of pure pops each precisely similar to some former pop.

As opposed to any theory of pure popping, therefore, this theory of creative process resolutely asserts the continuity of succession. The present moment inherits from the past. It receives its being in the very act by which the past passes into history, but it is a parent as soon as it is born. In its passing it presses on, and its creativeness is its forthcomingness. There is no such thing as creativeness without inheritance. Nevertheless, there is creativeness in the growing-point that we call the present.

Let us examine the problem from the side of inheritance. The maxim *Ex nihilo nihil, et in nihilum nil potest reverti* is obviously a guiding principle in all research. It means that we are always entitled to ask, or believe that we are always entitled to ask, "Where does it come from?" or "What does

[1] D. W. Gotschalk, *Structure and Reality*, p. 54.

Creation

it come out of?" when we set ourselves to investigate any matter of fact. That question however (its legitimacy admitted) may be almost as vague as it is wide. At its widest it would mean only that every matter-of-fact must have some kind of spatio-temporal continuity with earlier matter-of-fact. Nothing need be said in detail about the type of continuity. The statement, for instance, need not deny *actio in distans* if that occurs. It need not assert (as Kant[1] seems to have held) that wherever there seems to be an abrupt change of degree, all intermediate gradations must have been traversed. It need not deny action at a temporal distance, if that be the proper explanation of the continuity of a man's personal experience interrupted in sleep and in trances. All that need be meant is that there is some continuity somehow traceable.

Many of those who oppose the maxim in the name of what they call "creativity" do so in a special sense. They oppose the doctrine that substance is metaphysically as well as physically unalterable, ageless, unhistorical, and that all changes are changes, not of substance, but in the relations between substances, their distance, separation and combination. This they call contemptuously a theory of "mere reshuffling".

In such a doctrine the problem of the relations between stability and vicissitude is partitioned into two realms, the realm of abiding substance and the realm of changing relations. It seems to be quite clear that any such division yields no solution. What has to be sought is stability *in* change, not stability *and* change. The reference to relations does not help. There are no relationless substances. But there are changing substances. And if anything is unalterable, certain relations are. If for example A is earlier than B, it is so once and for all. On the other hand, it is easier to perceive the defect in this and in many other attempted solutions than to discern an efficient remedy.

To be consistent, I think, the theory of creative process must hold that anything that is said to persist, persists as a

[1] *Critique of Pure Reason*, "Anticipations of Perception"

flame persists. In the ordinary way we distinguish between the flame's persistence and the persistence of candle or wax which, for a time at least, do not seem to need renewal. According to the theory of creative process there is a relative, but there cannot be an ultimate distinction of this order. The candle, the wax and its chemical constituents are all renewed from moment to moment. They all go on. We speak of their stability when their renewal seems to involve no change of properties except the renewal itself. We speak of their mutation when their pattern seems to dissolve like the smoke messages from an aeroplane. But there is absolute becoming everywhere.

This view has certain striking resemblances to Descartes' theory of the continual creation, that is, of the continual re-creation of the world. It differs principally because it applies to God's existence as well as to the world's. The proof of his theory, Descartes said (*Principles*,[1] I, xxi) clearly appeared "when we consider the nature of time, or the duration of things; for this is of such a kind that its parts are not mutually dependent and never co-existent; and, accordingly, from the fact that we are now, it does not necessarily follow that we shall be a moment afterwards, unless some cause, viz. that which first produced us, shall, as it were, continually reproduce us, that is, conserve us". According to the theory I have sketched that which is actual at any moment begets a successor and, in that sense, dies to live. Present actuality, therefore, does of itself what according to Descartes, God had continually to do for the world, that is, renew its existence. Attend with Descartes to the nature of time. It always describes renewal. Therefore the endurance of temporal things is their continual renewal. If all existence is of this order, it is creatively self-perpetuating in a movement that is also inheritance.

---

[1] Veitch's translation. Cf. *Meditation*, III. A parallel may be found in the Mahommedan philosopher, Al Ashari (873–935). See Gilson, *The Unity of Philosophical Experience*, p. 45.

Creation

Let me treat, at somewhat greater length than I have already done, of the relation of such creativeness to novelty.

Cartesian re-creation need not imply more than a drab metaphysical "novelty". One dawn may be very like another dawn; yet each must be renewed. One pea may be very like another pea; yet each has to be re-created from moment to moment. The metaphysics of such renewal may be less adventurous than the periodical milking of a cow. It is compatible with the routine of the best-drilled event-patterns. Again there is nothing in the metaphysics of renewal to suggest the inevitability of creative advance, if the word "advance" implies improvement or betterment. There is pressing *on*, but not necessarily, with St. Paul, towards any mark at all.

What is ultimate is the continuity of process, and that in itself has no implications at all regarding the extent to which history does or does not repeat itself. The absence of such implications, however, has a certain emancipating power. It neither compels us to hold that the later must be similar to the earlier, nor compels us to assert the dissimilarity of the later. There must be a continuous history, but we are entitled to accept any pattern of continuity for which there is sufficient evidence. If, as I have suggested, the future (as we call it) is always indeterminate before it occurs, it follows that we can never be wholly certain, in advance, of what it will be. That circumstance has important metaphysical and theological consequences, but it does not forbid and it does not impede the guess-work regarding the future that we find in almanacs, and timetables and accept in a great host of scientific and practical affairs.

Consequently there is no need to deny the occurrence of "genuine novelty", although I should not care to try to define the phrase with precision. It means more than what I have called Cartesian re-creation. It means more than the occurrence of something somehow dissimilar from anything that ever occurred before; for that would be entirely consistent with the theory of "mere re-shuffling". Illustrations may be given

135

from the empirical fields that have already been surveyed at an earlier stage of this lecture—a work of artistic genius, the birth of a man, and the like. Each such event, in comparison with its antecedents, may be called a "genuine novelty", not merely in the sense that it surprises us—for the most venerable thing may do that—but also in a further sense that may have some importance and is not quite arbitrary. The theory of "emergence" may also illustrate the possibility. But as I say I should not like to try to define positively what this "genuine novelty" is.

I have now attempted to give a sketch of the metaphysical theory that all existence is in process and that creativeness is a necessary aspect of anything that is in process. This theory, as it seems to me, is immensely stronger than any attempt to scratch for creativeness in some special sense, or for "genuine novelty" of a striking kind, within the world. For my part I believe it to be true; but, as I have repeatedly observed, it is hostile to the theological view that the world proceeds from a God who does not himself proceed. The main discussion of this subject, however, must be postponed to the next lecture which will deal with "eternity". I should not anticipate that discussion, but I should like, in the rest of the present lecture, to develop the theological implications of the distinction between the theory that all actuality is creative as such, and the theory that worldly actuality is created by God. In doing so I am very well aware that the problems of time and vicissitude are the hardest in all metaphysics. If I speak dogmatically on this subject, and in a way that is distressing to many cherished religious beliefs, the reason, I am afraid, is rather to conceal than to deny my own diffidence. Diffident as I am, however, I have to express the conclusion towards which I incline.

The theory of the creativeness of all reality denies the legitimacy of trying to find an explanation of becoming itself, that is to say, it refuses to peer behind the scenes in order to find something that makes going go. It declares that process,

Creation

transition or becoming is wholly ultimate. In many expositions of Christian metaphysics the contrary is assumed. It is held that existence is ultimate while time or becoming is not. God's existence, the great "I am",[1] is said to be ultimate. There is no sense in going behind it, no sense in asking what originated it; but anything that *begins to be* (according to this theory) must be originated. The theory of creative actuality denies this conception *in toto*, and asserts that all existence is ultimately existential becoming. Consequently there is no sense in supposing that any existent, divine or secular, non-natural or natural, originates the process of becoming. There is no sense in the view that anything that passes, and consequently passes away, is, for that reason, dependent, contingent, unreal, self-insufficient; for all actuality is passage, passing away and also passing on. That applies to the world and to God. Both pass away. Both press on. Both are in the making, and if they were not in the making, would not exist. Both renew themselves. Both grow. Time does not, like an ever-rolling stream, bear all its sons away. On the contrary, what Isaac Watts called its "sons" are time's actuality and they bear time with them, although each is about to die.

Evidently, therefore, there is the greatest possible disparity between the theory that all that there is, including God, is in process just because it is in being, and the usual Christian doctrine that the world of becoming is generated or created by an ingenerable uncreated being who does not himself become or proceed but is "above" process as well as "above" the world (or shall I say? the underworld) of process. We may, however, raise the question whether if the being of God was a divine process, any meaning could be attached to the usual Christian doctrine.

According to a belief usually unsophisticated, the story of the creation of the world was something like this: God has existed from everlasting, without beginning. At a certain

[1] Cf. the Indian *Tat tvam asi*, That art Thou.

moment he produced the world (or the under-world) out of nothing. Since that moment the world has continued to exist along with its maker. At a later date it may be annihilated, and God may again be alone.

There seems to be no contradiction in this unsophisticated opinion. It is a possible piece of history and of prophecy. It frankly accepts the view that God is a temporal being existing from everlasting to everlasting, that is to say, is temporally infinite in a series open in both its directions. Within that series the world series occurs through God's creative acts. The world series has a beginning, and is therefore closed at one end, *a parte ante*. It might or might not be closed at the other end, *a parte post*. If it were closed at the other end it might still have an infinite number of members, although the number of such members might also be finite. But *God's* boundless temporal existence would be infinite.

Such a theory, so far as I can see, implies no contradiction. What is more, it could be held without internal absurdity by anyone who believed that all existence is temporal, growing and "creative". What would then be said would be something like this: There was a time when the growing God grew by himself. At a certain moment his growth began to include a relatively independent growth, namely the growth of the world. At some later moment there may cease to be any renewal of the world. Such a theory might indeed resemble the emanation theory much more closely than the creation theory, but the origin of the world might be so great a "novelty" that the special term "the creation" could legitimately be used to give it emphasis.

As I say, this view does not seem to me to contain any internal contradiction, and if it is unsophisticated I think we should say that some of the more usual objections to it are also rather childish. There is, for instance, no signal profundity in asking why God was idle for so long, and why he selected one moment rather than another for creating the world. If a man will talk in that vein, why should he suppose that God is idle when he isn't busy about world-making? Is that his only occupation? Would he have nothing to do if he were done with the world? And why should it be a reproach if he did not do anything? What is the virtue in a theory of God the hustler? Again, if we ask: "Why one moment rather than another?" it might be enough to reply: "Why *not* this moment as well as any other?" There would, on the hypothesis, be nothing *outside* God to make it wise or prudent for him to wait; but he might have ample reason within himself. It is true that we do not know what the reason is; but we don't know everything. If, finally, we ask: "Why should he have waited so long?" the question would not seem to be very penetrating. How do we know how long he waited? He needn't have waited for more than a jiffy. And when we complain of the length of the delay, what standards are we applying? *Why*, if the delay were long, should we say that it was *too* long?

For the most part, however, it is assumed that the "world" in all these arguments should be understood to mean the entire order of becoming, of time and of process, and that the cosmological and philosophical problem of creation concerns the "world" in that sense. If so, our recent contentions are most highly relevant. Process or becoming, we say, is absolute. If it had a beginning, it began by pure popping, that is to say it just began. It did not originate *from a source*, or proceed from anything that did not itself proceed. Since there is no need for assuming that the process of becoming itself began it is reasonable to interpret the absoluteness of becoming as including the presumption that the world had no beginning, and to affirm that any actual moment in the world-process belongs to a succession that inherits from the past and presses on towards the future. That is an admission of creativity but a denial of creation *ex nihilo* and also a denial of the world's creation *ab extra*.

Aristotle in *De Caelo* (A, 10, 280 a 28) censured Plato for having spoken in the *Timaeus* (52 d 4) as if the world had a

Theism and Cosmology

beginning but no end, and in one passage in the *Timaeus* Plato spoke about what happened "before" God made the universe (οὐρανός). As Aristotle's contemporary Xenocrates remarked, however, Plato also said in the *Timaeus* (38 b 6) that time came into being along with the universe. Commenting upon these matters, a distinguished modern writer, Mr. A. E. Taylor, says: "No sane man could be meant to be understood literally in maintaining at once that time and the world began together, and also that there was a state of things, which he proceeds to describe, *before* there was any world. The 'beginning' of the world at a certain date must be taken to be only part of the mythical imagery; the truth it is intended to convey is simply that the world we perceive by sense depends on a cause other than itself—God."[1]

Aristotle's comment may have been based upon the assumption that "time" referred always to the regular movements of the heavenly bodies. If so, Plato in his *Timaeus* (37 c) did speak of the "everlasting gods" in that sense, and called the world their "moving likeness" or, as Mr. Cornford translates, "a shrine brought into being"[2] for them. The more general question, however, is also the more important. That is the question whether the process of becoming (γένεσις) has its source in a being (οὐσία) that is not itself in process of becoming. If the latter were the truth it might be relatively unimportant whether we were dealing with myth or with philosophy. The contention, however it were expressed, would be the thing to debate.

To say that the world is the moving likeness of an eternal model[3] is a statement that may be interpreted in many ways. The meaning may be only that the regularity of astronomical movement within the world resembles in its constancy the timeless, because unchanging, relations that some philosophers regard as the peculiar province of "reason". According to Aristotle, "God fulfilled the perfection of the universe by

[1] *A Commentary on Plato's Timaeus*, p. 69.
[2] *Plato's Cosmology*, p. 97.
[3] *Timaeus*, 37 d.

Creation

making coming-to-be uninterrupted; for the greatest possible coherence would thus be secured to existence, because that coming-to-be should itself come-to-be perpetually is the closest approximation to eternal being".[1] If as I am maintaining "coming to be" is an attribute of all existence including divine existence it cannot be an image of a model that does not come to be. The creation *of becoming* cannot be an inferior replica of anything. The so-called model and the so-called replica, if they exist, must both come-to-be and press on. Both are "in the making". Their *Wesen* is *Wesensbildung*.

So far as I can discover from reading and from conversation, there are not very many people nowadays who believe that the world was literally created *ex nihilo*, but many people believe that the doctrine conveys profound truth in a figure that is bold and impressive. That also, as I have said, is among my difficulties. The figure, as it seems to me, is intended to symbolize the overlapping of coming-to-be by something greater that does not come-to-be. In that case it would be *only* a figure because the overlapping cannot strictly be temporal. These topics will be discussed in the next lecture which deals with eternity in its relation to time, and in a subsequent lecture which will deal, *inter alia*, with the problem of what are called "first causes". My principal trouble is that I cannot believe that coming-to-be is not the marrow and essence of all actuality. If it is, the figurative language mentioned above is not more helpful than literal language. The whole enterprise is mistaken.

I should like, however, to repeat the expressions of diffidence that I made before. However seriously one may take the problems of time and of time-like process, they have an evasive habit of destroying confidence. Even the least self-critical among philosophers must admit the fact. If theology sets about to belittle time—and it does so very often—its attempt easily arouses a strong and resonant sympathetic chord in nearly every heart and in a great many heads. I allow,

[1] *Physics*, 223 b 15. Trans. J. L. Stocks, *Time, Cause and Eternity*, p. 19.

therefore, that I am even more likely to be wrong in this than in most other difficult matters. In what I have said, however, I have the backing of much philosophy that is justly esteemed, and so have to ask for rather less forbearance than would otherwise be necessary.

ADDENDUM TO LECTURE IV

I may seem to have paid too little attention, in this lecture, to the questions *why* and *how* there was the creation of the universe; and similarly regarding its emanation. The reason was that I was trying to show that existence itself, even God's, must be held to be in a certain sense a creative process, and that *such* creation or emanation is ultimate.

Whatever Christian theologians may think of the positive part of this argument, my omissions concerning *how* and *why* should not surprise them. The general and, I believe, the orthodox opinion among them is that the manner of creation, and the reasons for it, must remain a mystery that philosophy may indeed accept (for it contains, we are told, no contradiction) but that philosophers and theologians cannot explain.

Accordingly we need not even smile if mythopoeic attempts to picture the creation are often feeble. I need not refer more fully than I have done to the Great Egg, to pre-cosmic rain, to the "primordial germ", to rays from Hari's forehead, to cosmic emergence from a lotus on the navel of the Unmanifest, or to other such instances of the imagination of the East.

There may seem to be more of philosophy in the doctrines, maintained at many times and in many places, of a cosmogony which displays the emergence of the cosmos from a primitive formlessness, of a Form or unity that required and produced multiplicity, and the like.

The first of these opinions, namely that formlessness acquired form, perhaps retaining it, perhaps falling back into formlessness, perhaps proceeding from formlessness to form and back to formlessness in endless cycles, has been frequent

Creation

in the East and elsewhere. It is ontological if it speaks of the process of the Undeveloped into the Developed and back again. It is epistemological rather than ontological if the doctrine asserts that the Unmanifest becomes manifest (and then returns to its mysterious obscurity). In the last case we have, among others, the extreme doctrine that what we call the universe is the *Maya* with which Brahman deceives himself.

From China in the Huai-Nan-Tzŭ we have the following: "When Heaven and Earth did not yet have form, there was a state of amorphous formlessness. Therefore this is termed the Great Beginning. This great Beginning produced an empty extensiveness, and this empty extensiveness produced the cosmos. The cosmos produced the primal fluid which had its limits. That which was clear and light collected to form Heaven. That which was heavy and turbid congealed to produce Earth."[1] Hence *yin* and *yang* (acting as male and female principles), the seasons and the "myriad things".

The general theory that Unity is the prius of a complex world is also familiar in philosophy. One of the simplest forms of it occurs in Indian philosophy. Brahman said "May I be many". It must always remain a mystery however why the Unity should divide itself, whether it is a primordially undifferentiated unity or a simplicity in which all differences are reconciled. Attempts both in China and in India, as well as in Islam or in the pages of the Pseudo-Dionysius, cannot make the theory plausible. Such a Unity, or Unitarianism, is inevitably embarrassed by the world. For the world must be very unlike God who alone is the truth and the reality. Why then should there be a world, even if it exists to be redeemed (or deified)? Some of the Sufis say that the reason has to do with knowledge. As it has been said: "Before all things, before the creation, God in his unity was holding an ineffable discourse with himself, and contemplating the splendour of his essence in itself. Then God by his essence in his essence desired to project out of himself his supreme joy, that Love

[1] *Chinese Philosophy*, by Fun Yu-Lang, trans. D. Bodde, p. 396.

Theism and Cosmology

in aloneness, that he might behold and speak to it. He looked in eternity and brought forth from non-existence an image, an image of Himself, endowed with all his attributes and all his names, Adam"[1]—or, by other accounts, the archetype of Mahommed.

It is plain the God's desire for an audience or for some sort of self-mirroring is not readily intelligible.

The Logos theory does not seem to have any natural affinity with the doctrines either of creation or of emanation, unless indeed the Logos is regarded as a magical name, in which case the Indian thinker who said that Brahman produced the cosmos by his own magic would have been very profound. If by the Logos is meant a logical essence (or the formula for such an essence) there can be no sure foundation for the belief that *such* essences produce physical or psychical existents. Things *have* but are not *made by* such essences. Similarly things cannot emanate from such essences, and they are "incarnations" of the essences only in the sense that they *have* them. The same type of argument would hold, *mutatis mutandis*, of the idealistic as opposed to the realistic forms of this doctrine, where the "essences" are interpreted as noëta and the "things" as sensa.

According to many theologians, not all of them Christians, the creation or emanation or enhylation (incarnation) of the cosmos is to be regarded as due to God's love. It is an overflow from divinity, a voluntary self-emptying or kenosis of God's boundless bounty. Such a view is unintelligible unless Perfection is made better by such an overflow, and that is very hard to believe. It is not a denial of a growing God.

I note with interest that a recent theologian, Mr. Bertocci, appears to be satisfied with the view that what has happened is just that God had first of all to overcome obstacles within his own nature. These being surmounted, he was then able to produce the universe and he did so.[2]

[1] Quoted by R. A. Nicholson, *The Idea of Personality in Sufism*, p. 29.
[2] *The Empirical Argument for God*, p. 113.

V

Eternity

In the last lecture we considered temporal process in relation to the cosmological theory of creation. At its close we seemed to be impelled, gradually but very firmly, towards a discussion of the relations between time and eternity. That is a bigger and a deeper theme. Devout persons and the mystics in all recorded ages have sought and hoped for an escape from vicissitude, and for a type of security that seems absent from the fading pageant of the world. God for them is stability itself. If that be a dream it has at least been among the most moving of human experiences. If it has metaphysical foundations that cannot be shaken, its importance, from every aspect, cannot be gainsaid.

I shall divide the discussion into three parts. In the first I shall deal with the stubbornness of time and of change in metaphysics. In the second I shall discuss the meanings that may be given to eternity. In the third I shall consider change and eternity together, with special reference to theism.

Consider, firstly, the stubbornness of time and of change. I shall try to show (1) that even the illusion of change itself implies change, and (2) that what is called "relative" change implies change *simpliciter*.

(1) I do not think the proof is hard. Take the supposed illusion of change. This must mean that something, X, appears to change when in fact it does not change at all. That may be true about X; but how could the illusion occur unless there were some change somewhere? If there is no change in X, there must be a change in the deluded mind that contemplates X. The illusion of a change is actually a changing illusion.

Thus the illusion of change implies the reality of *some* change. Change, therefore, is invincible in its stubbornness; for no one can deny the appearance of change.

(2) Take, again, what is called a "relative change". It may be only simpler and not truer to say that the train approaches Glasgow, than that Glasgow approaches the train; but in either case there definitely is a change. We can reckon Glasgow as fixed, or we can reckon the train as fixed; but we cannot reckon them both as fixed, and also bring their relative change into the reckoning.

In general we describe change in terms of substance and of time. *Something*, we hold, changes, that is to say, has properties at one time that it does not have at another time. This language, however, may be rather too elaborate for the simplicity of change.

Consider "substance". The change, we say, must occur in the same thing, or in a thing's relations to other things. If not, there would merely be a difference of events; and different events need not, as such, constitute a change. In any ordinary sense of "substance", however, it is not strictly necessary to hold that all change must occur within a substance or between substances. For we can assert, without any philosophical impropriety, that a situation has changed or that a tendency has changed or that the weather has changed, although situation, tendency, or weather are not, in the ordinary sense, substances. All that is essential is that there should be a certain continuity of succession in what might be called the change-over.

Consider "time". According to Aristotle time was "the number of movement in respect of before and after",[1] and Aristotle passed rather quickly from the conception of time as a number to the conception of time as something numbered. Similarly we are accustomed to speak of time either as the measure of physical change or as that which is measured in physical changes. In other words, we equate it with chrono-

[1] *Physics*, 219 b 1–3.

Eternity

metry. This conception is obviously very elaborate and disputable in many respects. It may appear, indeed, to be rather simple in Newtonian physics where an *absolute* and *even* flow of time was asserted, but *such* "time" was admitted to be impossible to correlate quite precisely with most at least of the observable movements on which all scientific measurement of physical changes depends. In modern relativity-theory, with its local times and its formulae for correlating these, absolute measurable time is abandoned. That, in its way, may be a gain in simplicity. In other ways, however, it is immensely complicated, and often it is very hard to believe that we are not acquainted with something simpler which, if it be not "time", is at any rate incorrigibly time-like.

I propose, then, to examine our simpler time-like experiences with some assiduity.

Take, for instance, a dream. The dream, we may admit, occurs during a definite period of clock-time, but the dreamer need not have any dream-clocks, and he is exiled, during his sleep, from the ordinary workaday clocks. Yet he may and does experience succession in his dream. Succession, consequently, may occur and be noticed when clock-time cannot be noticed. It has quite a good meaning when its expression in terms of public measured time is altogether impossible for the man who experiences it.

Of time in general it would commonly have been said (before men began to submerge it in space-time) that there was just *one* time, an objective time series. Of this series we could say that it had direction since it proceeded from earlier to later, that earlier and later were closely connected with past, present and future, and that the order from earlier to later was transitive and fixed. [It was transitive because if A is earlier than B, and B than C, A must be earlier than C. It was fixed because B, if it is later than A, can never become earlier than A.] This single order of time, it was said, involved relations of temporal distance and had no timeless gaps. It also, as Newton said, "flowed evenly". Further questions about time

Theism and Cosmology

were whether it could have beginning or end, and whether reality was "in" time or time "in" reality. In any of its senses time is more elaborate than simple change.

If we compare these characteristics with the timelikeness of dreams we obtain, I think, the following results: In the dream there is succession, and so there must be earlier and later as well as past, present and future. The dream process is transitive and has a fixed order in its intrinsic seeming, and it contains relations of time-like distance. In other respects, however, it would appear to be structurally inchoate. We can scarcely ask significantly whether the time-like process in the dream flows evenly or has gaps in it. Again, the dreamer, while he dreams, need have no inkling of position in public time. The dream, in its mere seeming, would appear to have a beginning and an end; and I do not see how we could argue the question whether the dream is in dream-time or dream-time is in the dream.

Even in a dream, however, succession is absolute fact. If we perceive it, there it is, whatever may be true about the rest of the universe. It is authentic fact, and cannot be dismissed as fancy, even when it occurs during the process of fancying. Being authentic it must really have the properties it implies, namely that in it the earlier precedes the later.

To affirm that succession is absolute, however, is not to affirm the metaphysical necessity for Newtonian absolute time. There are no landmarks in so-called absolute time, as Newtonians themselves admit. Therefore measurable physical time has to be measured dynamically with reference to the swing of the pendulum or the journeys of sun and stars, or perhaps in a different fashion, as Professors Milne and Dirac suggest to-day, by such atomic changes as the rate of disintegration of radium. Such measurement depends upon the observation of physical movement, and upon assumptions regarding the uniformity of physical movements. It may require a thoroughgoing inter-relativity of "space" and of "time" when it is expressed in physical equations.

Eternity

The simple experience of succession is not a witness in this matter. As we have seen, it is independent of clocks, and sometimes we may not be able to correlate it with anything else in the universe. Dreams may be instances. On the other hand, the fact that we experience absolute succession, although it may have very little to show regarding the technique and the useful conventions of physical chronometry, has something very significant to show about the metaphysics of this matter, that is, about the naked truth of the affair. It is quite impossible that if B really and absolutely does succeed A, it can also be simultaneous with A or precede it. This obvious truth need not, indeed, occasion much excitement among the physicists. They are concerned with measurable chronometry, and may surrender dreams to philosophical dreamers; but even the dreamers have something relevant to say.

To continue with this simple illustration of the dream, it would be very rash to affirm that because the dream is evidence of absolute succession, it is evidence of an absolute *rate* of succession, or of the "even flow" of such succession. Yet without such evidence we should (perhaps) have been introduced only to what was time-like and not to time. Definite relations of earlier and later would not be time-relations, in the ordinary full sense, if there were no discernible equality of interval. If dream or other simple time-like experiences provide either no evidence or hopelessly inadequate evidence on this point, their silence would be very significant indeed.

On the other hand, there are dangers in exaggerated caution. In illustrating the simple experience of succession from our dreams I have purposely chosen to illustrate from a type of experience which, by common consent, has the *least* information to give about real existence. If something relevant to that issue can be extracted even from dreams, the result may be of great importance precisely because its basis has so little promise. Other simple experiences in waking life, in vigilant perception, and in careful introspection, might then be very promising indeed; and surely these are ultimately our principal

Theism and Cosmology

resource. If rate of change is to be measured chronometrically, the measurement has to be based (has it not?) upon the observation of succession, that is, of concrete not of empty passage. Therefore our simpler experiences of passage yield the ultimate data for our physics, at any rate in terms of the primary assumptions of contemporary physics. If we had no experience of the over-lapping of successions and of their approximate steadiness we should have nothing at all to go upon. We might, it is true, assume that pendulums and stars moved uniformly and so might construct a consistent but quite arbitrary system of physical chronometry. If, however, we have the right to assert that such a system is not arbitrary, the reason must be that we can form estimates of time-like distance and interval, not wholly unreliable although, it may be, very imperfect.

Certainly we have to show a proper circumspection. Everyone knows that our subjective time-like estimates may differ most appreciably from clock-time. "I'll tell you", said Rosalind in *As You Like It*, "who time ambles withal, who time trots withal, who time gallops withal, and who he stands still withal." When we are bored a shortish clock-period seems long; when we are agog with excitement a clock-hour may seem to have lasted for a very few minutes. Again, it may reasonably be held, as by De Noüy[1] and others, that our subjective time-estimates largely depend upon the physiological time-piece that every man has, viz. his own body, although we may not usually be aware of the circumstance. We may estimate half-consciously from our regular breathing (when we are not catching a bus), by our digestive clocks and generally by the repair of tissue. True, there is no reason to suppose that our lung-clocks or our stomach-clocks are peculiarly reliable time-keepers; but they are not wholly ineffective.

These difficulties, however, are subsidiary. If a sundial is a better clock than one's growing hunger, the reason is that a

[1] *Biological Time.*

Eternity

wider consilience of world-testimony can be obtained by trusting sundials and their kind than by trusting one's hunger. This wider consilience in its turn, however, is a consilience regarding rates of change within the world, and that must imply, in the end, accordance with our actual experience of rate of change. Such experience may be very erratic and may often deceive, but unless it is trustworthy *au fond*, and unless its mistakes are corrigible by more patient observations of the same order, neither world-chronometry nor extensive local times could be more than a beautifully finished piece of convention.

These latter considerations, although less important than the former, are evidence of the stubbornness of change and of the time-like in a simple and final sense. Another consideration treads close upon their heels, and is of the utmost moment.

In the normal way we think of time as forming a single series. There is, we think, a single universal time-order, whether or not it is possible to assign position and distance in absolute time. Hence we find difficulty in the conception of local times (in the sense asserted by relativists) although we may readily be forced to admit that if there has to be communication by signals which take time to travel, complications are bound to arise. We may even become reconciled to the notion that *measurable* motion involves the inescapable and integral inter-dependence of "spatial" and of "temporal" factors.

If, however, it be a metaphysical truth that nothing is determinate before it occurs, it would seem to be altogether impossible that there could be anything relative about such a matter. How could it be supposed, except for instant dismissal of the idea, that anything could *be* from one point of reference, and yet, from another point of reference, would not exist at all? That sort of language might indeed be used about "epistemological objects" if there were any; but we have seen that there can be no such things. Failing these monstrosities,

Theism and Cosmology

if actuality be always present actuality, there can be no occasion or excuse for holding that what is actual in one system might not be actual in another system. For actuality is absolute *ex hypothesi*, and cannot be limited, either expressly or covertly, to any particular system or standard of reference. We may not be able, indeed, on the basis of simple experience, to effect satisfactory correlations of co-existence between our own observations and those of other observers on the earth, to say nothing of hypothetical observers light-aeons away; and there may be much illusion in our own private time-like series. That, however, does not affect the question. If the present has properties that neither past nor future can have, nothing can alter the circumstance. The past, being determinate, may indeed be reported belatedly, but the future, being indeterminate before it occurs (*id est*, nothing definite) cannot be reported as something definite. Signals might perhaps be capable of travelling instantaneously; but they could not travel in less than no time.

Accordingly, if the view suggested in the last lecture be true, viz. that the past at any moment is determinate but impotent, the present determinate and powerful, the future indeterminate, therefore nothing real,[1] therefore impotent, it follows that this is an absolute statement for all existence worldly or over-worldly. I shall return to this question later; but now, having explained why time or the time-like is so very stubborn as well as mere change or succession,[2] I have to consider why

---

[1] I agree with most that Mr. Broad has written in *Scientific Thought*, Part I, chap. ii, even if Mr. Broad no longer agrees.

[2] Regarding the relations of these, I should say, generally, that change implies succession, if, indeed, it is not identical with succession: that succession is time-like when there is distance and order in it, even vaguely as in a dream: that "time" is of the same order as what I have called the "time-like", but more elaborate and more exact in respect of distance and of orderliness; and that chronometry or the measurement of time has to be distinguished from the time which is measured. Questions about private succession in its relation to public time, the extent to which chronometry depends upon physical measurement, and the like, are subordinate, however important they may be.

certain philosophers are prepared, nevertheless, to renounce its reality.

Their principal argument is that contradictory assertions cannot both be true of the same thing. There is no contradiction, they admit, in the series of earlier and later. If X is earlier than Y it is always so, provided that both have occurred. It can never become later than Y or simultaneous with Y. They maintain, however, that the series of earlier and later has to be generated from the experience of pastness, presentness and futurity. It would have no meaning, they say, if it were not so generated. Past, present and future, however, are incompatible characteristics. Therefore nothing can possess them all; yet if change were real, *every* event in a temporal or time-like series (except the first and the last) would "have" (tenselessly) all three characteristics. What we call the present at any moment was future at an earlier moment and will be past at a later moment. If we suggest that the tense in these statements removes the contradiction we are deluding ourselves (they say). We employ tense to express position in the series of earlier and later, but that series is itself generated from past, present and future.

To this the obvious and (I suggest) the sufficient answer is that there would not be change or succession *unless* the present passed away and pressed on towards the future. The present could not be a changing event if it were always a present event. To suppose anything else would be to contradict the conception of change, and it can never be legitimate to regard an implication of change as the denial of change. Any occurrence is actual when it is present, and then it is about to pass away and about to give place to a new occurrence. That is a description of change or succession itself, and is a consistent description of it.

Another objection to the reality of change concerns the dimensions of the present moment. Such a moment, it is objected, is only a mathematical point having neither temporal breadth nor temporal thickness. Such a point could not be experienced. It is frankly incredible, we are told, that it could be the only bearer of growth and of power.

In this matter, several modern British philosophers, following William James,[1] who followed E. R. Clay, adopt the conception of the "specious", perceptible or apparent present. Such a specious present, they aver, is a perceptible slab of passage, the shortest time-like stretch that we are able to perceive when we say "This is happening now". It is sometimes described as the span of transition that is perceived in a single act of attention. If so, it might be of variable length according to the perceiver's vigilance, but its duration is commonly said to be fairly constant and to be about two-thirds of a second as measured by a metronome.

So interpreted, the "specious present" seems to me to solve no philosophical problem whatsoever. It is a stretch of transition. If it lasts for seven-tenths, or for seven-hundredths or for seven-thousandths of a second, the fact remains that its beginning, middle and end are not coincident, and that its later phases do not exist when its earlier phases occur. Philosophical sins are not condoned by the explanation that they are very little ones. The present is a name for the moving point of actuality in any transition, and if the phrase "the specious present" is intended to describe an apparent arrest of transition it is not even speciously accurate. For the transition does *not* appear to be arrested. The most that the phrase can describe is the fact that we perceive, not a durationless "present", but a short transition within which the present has been advancing. In other words, any perception that contains the present as its later boundary is partly retrospective, but can never anticipate futurity (except, of course, by conjecture). This circumstance may have technical importance for psychologists who usually define "perception" as something "present" and therefore distinguish it from memory as well as from expectation. It might also tend to show that instead of "perceiving" the present *per se* we ascertain it by reflection upon the implications of transition. But why not?

[1] *Principles of Psychology*, I, 609.

Eternity

Such reflective analysis encounters the puzzles of infinite divisibility. It is not, indeed, wholly certain that a stretch of perceptible transition contains no gaps and no next terms. If it were not continuous in the strictest sense it might nevertheless be analogous to the blind spot where no gap is perceived. Our time-acquaintance might be stippled with temporal blind-spots. But we could not accept such a view without renouncing the continuity of process, and there is nothing to compel the renunciation. The question therefore is whether infinite divisibility should cause any dismay. It seems to me that it should not. It implies no contradiction. Any line of finite length can be cut at an infinite number of positions. Therefore it contains an infinite number of positions; and although a mere aggregate of points could not compose a line, they *are* the line when they are serried in a certain relation. So here. A transition is not made up of an aggregate of present positions. On the contrary, any stretch of time contains an infinite number of *successively* present constituents. The stretch as a whole is never literally "present". Yet each such "present" is the point of actuality.

I submit, then, that change and the time-like are invincibly stubborn in metaphysics, *id est*, in ultimate fact. No objections are strong enough to overwhelm them, and that is scarcely surprising in view of the considerations with which this lecture began. For even the illusion of change implies the reality of some change somewhere.

Let us pass, then, to the second part of this lecture, and consider the meaning of "eternity".

The most usual meaning is frankly temporal. That is eternal which endures for ever, is everlasting, is permanent *in saecula saeculorum*. Such everlastingness implies unlimited endurance *a parte ante* and *a parte post*. It is illustrated inadequately by the everlasting hills, or has been so illustrated up to the present. It might be illustrated even more adequately by atoms or other supposedly indestructible substances; for these have often been regarded as strictly imperishable, and at any rate seem

Theism and Cosmology

to be very different indeed from what von Hügel[1] called the "slush" and "mere racket" of change.

It is plain that such immutable permanence need not conflict with the time process and need not be supposed to enlarge that conception. Certain distinctions have indeed to be noted, but these would have to be remarked in any case.

If the correct analysis of process be continual renewal in succession, anything that we call permanent has been and (we suppose) will be continually renewed. That implies passing away and coming to be, but if we say that permanent things do not perish, our statement would be incomplete if we did not explain what we meant by a "permanent" thing. A "permanent thing" in the sense that is relevant here has sometimes been called a "continuant", and such a "continuant" would describe a certain unity in sequence. The hills have continued for a long time with a recognizable continuous contour and with other perceptibly stable properties. In calling them "continuants" or "continuing things" we apply the name to the whole successive unity as long as it has lasted, that is, so long as it has been successively renewed. Therefore we contrast such continuants with, say, a flash of lightning. The latter when it passes away is not renewed, or at any rate is renewed intermittently and usually in a different place. The former pass away by the mere fact that they pass *on*, but they are continuously renewed and for that reason are called continuants. Since even the hills are not wholly immutable there will probably come a time when we should deny this recognizable obvious continuity to *such* continuants. We should then say that such a continuant had passed away altogether. *It* would not have been renewed, although its successors may well have had another and perhaps a less obvious sort of continuity with their predecessors. Our standards in these matters, it is true, are often rather arbitrary and are sometimes quite superficial; but that, if it be an objection, is only an

[1] *Selected Letters*, p. 364.

Eternity

objection to the sufficiency of our empirical criteria of thinghood.

In the extreme case this conception would apply to continuants (if there were any of them) which changed in no respect except in the respect of being renewed, and it would apply rather more plausibly to the atoms of the ancients whose renewal was supposed to be immutable in all respects except change of place. For the most part, however, we do not hold, and there is no reason why we should hold, that most "continuants" do not vary in a great many respects. A living body, for instance, grows old, not in the sense that it has continued in the ageless way of an atom, but in the sense of greying hair, and stiffening joints and length of tooth. Descriptively that is what happens in the case of certain "continuants", and although we may make the continuity easier to grasp by holding that the development of ageing continuants depends upon the persistence of constituents that show less radical alterations, it seems unlikely, at the end of the day, that there is any way of getting behind the descriptive fact of continuing thinghood. If so, the presumption is that there *is* nothing behind it.

Accordingly, if it be maintained that either God or the world is eternal in this sense of the word, what is asserted is that the world or God is a continuant whose existence is successively renewed without beginning or end. There is nothing in this view to forbid the conception of a growing God (provided that he always exists) or, of a world in the making (provided that there always is that continuant). Nothing is implied regarding the type or degree of unity that such a continuant contains, except that we have at least as good a right to speak of continuity of succession in its case, as we have in the case of sempiternal hills or of immutable atoms or (some would say) of pre- and post-existent "immortal" human souls. If there be any other sense in which a "continuant" may be said to persist immutably, that sense would also be applicable to God or to the world; but I see no need for postulating any such sense if eternity mean endless duration.

Theism and Cosmology

A second meaning of "eternity" is sometimes expressed in the phrase "an eternal present" or "an eternal now". "Aeternitas est simul et semel", Burman said in his conversations with Descartes.[1] The question in that case is whether such phrases describe anything that is actual.

The phrase would present little serious difficulty to those who believe in the "specious present". That, as we saw, may plausibly be described as the span of transition that can be grasped in a single act of attention, provided that the time-span so grasped appears to be continuous with the act of attending and does not, as in memory, appear to be the object of retro-attention. Such a "specious present" might vary in its clock-time duration. Under normal circumstances it is probably very different in man and in a dragon-fly. In God it might be infinitely longer than in man, and perhaps eternal *a parte ante*. On the other hand, if the future be nothing actual or determinate before it occurs there would be nothing actual or determinate to be included in any specious present, even God's, before its actual occurrence.

We, however, have denied the existence of the "specious present", holding that it is not a *mere* present, even speciously. It is experienced, we have said, as a transition in which there is passing away and coming-to-be, "just going" and *now*.[2] If, however, there seemed to be an arrest of passage in which there was no experience of waning or of waxing, there might then be some excuse for speaking of a "timeless" *now*, a *nunc stans*. It might be called an apparent *now* because it had or seemed to have the vividness and freshness of the present; and it might be called apparently timeless, because it seemed in our experience to contain no sense of transition at all, not even the sense of mere endurance without other change. That may be the negative side of the ecstasy of union with deity

---

[1] *Entretiens*, ed. Adam, p. 10.
[2] Dean Inge suggests (*God and the Astronomers*, p. 85) that the "specious present" is "present" only in the sense that it *contains* the present as in the phrase "the present year".

Eternity

that certain mystics believe they have experienced, although but rarely.

The mystics attempt to express what, according to their own theory, is ineffable. Hence they tend to use what are sometimes called "dynamic" or "transessentiated" negatives. What St. John of the Cross called "dispossession of self", was at the same time a suspiration with deity. God, according to Meister Eckhart, was "the still wilderness where no one is at home". According to Tauler, Eckhart's pupil, God was "pure Being, a waste of calm seclusion". "To be empty of everything created," said Eckhart in another passage, "that is to be full of God." The Haplosis or single-entitiedness of certain Neo-Platonists was of the same order, and so was the *Entwerden* of the German mystics. It has also been prominent in the wisdom of the East, as in the Upanishads, where even dreams were believed to give only a partial liberation from the pressure of daily life, and deep dreamless sleep was the only condition in which the emancipation could be fully accomplished.

Therefore, since they believed in the transessentiation of time, it is only to be expected that many mystics should sincerely believe that they had attained such a state, should regret that they had to describe it in ways that seemed to be but were not effable, and should enrich the literature of devotion by their attempts to describe what could not be set down in syntax. Natural theologians, however, may be permitted to doubt. The apparent arrest of temporal transition is not evidence of its actual arrest, however convincing such an appearance may be to the mystic who experiences it. That is plain because such an one has to say that he became eternal for a space, but not for long. He may think the explanation paltry, but he cannot think it wholly false. For the most part he has to say, like some Buddhists, that "the mind is perfumed by trances". In short, he has to say that he is sometimes "eternal" and sometimes not—which is frankly absurd. More generally the vigorous protest of Thomas Hobbes in his *Leviathan* cannot be disregarded. "For the meaning of eternity," Hobbes said, when he spoke of vain philosophy, "they will not have it to be an endless succession of time; for then they should not be able to render a reason how God's will, and preordaining of things to come, should not be before his prescience of the same, as the efficient cause before the effect, or agent before the action; nor of many other of their bold opinions concerning the incomprehensible nature of God. But they will teach us, that eternity is the standing still of the present time, a *nunc stans* as the schools call it; which neither they, nor any else understand, no more than they would a *hic stans* for an infinite greatness of place."[1]

A third meaning of "eternity" has to do with the timelessness of truths, of general conceptions, and the like.

Of truths we say correctly "Once true, ever after true". All truth is in that sense inconcussible and exempt from vicissitude. It achieves what Mr. Whitehead calls "objective immortality", even if the truth be about temporal events that have occurred, and about their temporality. If it be true that X occurred, that truth cannot afterwards be altered. One remembers the gourmet's joyful boast, "Fate cannot touch me, I have dined to-day". Fate may touch and destroy the gourmet's memory, but it cannot destroy the fact of his triumph. The achievement, like all achievements, is ultimate in the relevant sense.

Truths, therefore, that describe present and past (including truths that describe their temporal relationships) are, because true, timelessly so. Of the future, if the view I am defending is correct, we can speak hypothetically only. If to-morrow comes it will either rain or not rain. If it is fine the pavements will be dry. If the future be indeterminate before it occurs—I know I am speaking *Hibernicé*—no truth can determine it completely before it occurs. Therefore I have not said "Once true, always true", for that statement might be taken to mean that what we now know to be true, for example that Caesar loved Cleopatra, was true before Caesar met Cleopatra; and

[1] Chap. 46.

Eternity

I am not asserting any such proposition. I have said instead "once true, ever *after* true". Truth must refer to what is real. So referring, it is immutable; again in the relevant sense.

Similar remarks should be made about the eternity of concepts; for concepts themselves are timeless or unchangeable. This applies even to the concept of time or to the concept of change. If by reason of much philosophy we come to alter our conception of time or of change, we may or may not be wise, but the meaning of the statement is that we come to use different concepts to describe these realities or what we suppose to be such. The concepts do not change. *We* choose different concepts to express what we mean; and the facts may justify us.

Whatever is actual, then, has certain timeless properties; for there is nothing actual of which no truth holds or to which no concepts apply. That, however, would be true of the wildest flux. If, in such a flux, there were the colour orange, its hue would be between the hues of red and of yellow. If there were a circle, its radii would be equal. There is nothing here to cause surprise, and there is nothing to compel the inference that the possession of timeless properties warrants an existential distinction between eternal and temporal things. The concept of light does not itself shine or the concept of weight depress a balance. It does not even depress the concept of a balance; for such language is meaningless. No one, surely, is entitled to infer that because the concept of heaviness is not heavy, therefore, in all heavy things, there must be a part that is not heavy at all, but completely imponderable.

Summing up this discussion of the meanings of "eternity", therefore, we have to say that eternity in the sense of everlastingness is not non-temporal, that the apparent arrest of temporality in a standing *now* seems to be illusory (if, indeed, there is such an appearance) and that the possession of timeless properties is no sort of evidence for the existence of timeless things. These three meanings of "eternity" seem to me to be the only important ones; and so I turn to the third division of

Theism and Cosmology

this lecture, the relation between change and eternity with special reference to theism.

I shall begin by asserting three propositions. I do not see my way to deny any of them.

(1) The first is that changes in Y cannot wholly depend upon X unless X itself changes in some way. The mere fact of determining a change implies a change on the part of the determinant as well as a change on the part of the determined. No doubt there are many instances of what we call one-sided dependence in which one party does all the giving, and the other party does all the receiving. But the giving is a change on the part of the giver. It is a different fact from not giving, and it pertains to the giver's history. The concomitance of variations may not be a sufficient proof of causal connection; but a variation in Y that is inconcomitant with *any* variation in X is proof of the absence of complete causal dependence of Y upon X.

I may illustrate the point by referring briefly to Mr. Paton's erudite and illuminating work upon Kant. "Because of the nature of our mind," he says,[1] "things must appear to us as spatial and temporal; but it is because of the character of the thing-in-itself that we see one object as round and another as square." This view seems to me to be untenable when it is asserted generally about change. According to one interpretation of Kant's theory, things-in-themselves (or, in theistic language, God) neither change nor don't change. According to another interpretation they simply don't change. On either supposition it appears to me to be impossible that change in the phenomenal world should be a *phenomenon bene fundatum*, explicable in terms of its specific differences by distinctions in ultimate reality that are not themselves changes. Nothing but a variation can be the ground of a variation, and that will also be true, say, of rapid or of slow variations, a close parallel to the "round" and "square" in Mr. Paton's statement.

(2) My second proposition is this: If X and Y are co-

[1] *Kant's Metaphysic of Experience*, II, 417.

Eternity

existent entities, then if one of them be temporal, the other must also be temporal. Here, it may be objected, there is a play upon the word "co-exists". The "togetherness" is assumed to be a "togetherness" in time, although there might be togetherness in some non-temporal sense. If so, it might be suggested that one of the partners need not be temporal although the other was temporal. What I am asserting, then, is that there is no intelligible sense in which two *existent* entities, one of which is temporal, can be "together" unless both are temporal. A temporal existent may indeed have timeless properties, but as it seems to me, *two distinct things* cannot be co-existent or co-eternal unless they are bi-laterally temporal. If the world, let us say, endures so long as it is dependent on God, then God also must endure so long as the world depends on him.

(3) My third proposition has already received some attention in these lectures. Any attempt to explain change by an effective union of a changing with an unchanging substance or set of substances, at the best postpones an inevitable internal discrepancy. The two can neither lie down together nor get up together.

Therefore, if change be invincible—and we have seen that it is—there is no possibility of effective partnership or of grounded dependence between a changeless and a changing existent. The implications of this statement are quite general and apply to the relation between a divine unchanging over-world and a secular changing under-world as much as to any other reputed instance. We must hold, without limiting the statement in any way, that the problem of the relations between change and permanence can never be solved by an attempted happy union between things that change altogether and all the time, on the one hand, and, on the other hand, things that don't change at all. Therefore, change being invincible, the problem must be solved in another way, and there can be little doubt that that way is the way of Heracleitus, the father of the Logos doctrine in the West who further maintained that the Logos was eternal. If we look for permanence we must look for permanence *in* change, not for permanence *and* change. That quest is theological as much as secular. God's permanence must also be a permanence in change. Kant's statement that "only the permanent changes", if it be true, must be interpreted in the Heraclitean way. Passage or transition is ultimate in all existence. Everything that exists *becomes*, and its permanence or persistence consists in some identity of pattern as it is successively renewed. In the language of the scholastics there is no ultimate metaphysical distinction between *secundum fieri* and *secundum esse*.

Let us try to consider some of the principal ways in which there may be permanence or immutability *in* change or succession, and the relations of such immutability to theism.

One obvious view is the simple belief that, up to the present, there have been constant patterns, laws and regularities in the passage of reality, and that, in all probability, this constancy of pattern will be renewed as the process of reality goes on. This pattern, according to theists, is divine, according to their opponents it is secular. For both of them, however, the pattern should be a pattern *of change*, and not an eternal model of which changing events are some sort of image or dependency. Neither a divine nor a secular pattern would be Alpha and Omega in one. In respect of the nature of the pattern there might be no difference between Alpha and Omega. The same pattern might always be renewed. In respect of actuality there would be succession from moment to moment. If God be the All he is not simply the pattern. He is the patterned succession, and the same would be true, *mutatis mutandis*, if he were supreme but not the All.

Another view of permanence, not necessarily irreconcilable with the former, seems to be favoured by the supporters of M. Bergson, and has, I believe, the countenance of the master, although I will not here go into the exegesis of his philosophy. It is, to be brief, that the present always contains the whole of the past, by a sort of intussusception or absorption, although

Eternity

each successive present moment enriches the store of being. This does not mean simply, as Mr. Broad would say,[1] that the "sum total of reality", i.e. past and present, grows larger every moment. It means that each successive moment literally contains the totality of its antecedents. There may be some feeble empirical indications of such a theory in living things and in the psyche; but even if in living things and in the psyche passing *away* had to be accepted with the same faithful matter-of-factness as passing *on*, the metaphysical conservation of the preterite, in a universal sense, might still be a tenable theory.

Such a theory would be comforting if the acquisitions were good, and depressing if they were bad. In a certain sense, however, the theory might be said to guarantee a certain type of security; and it is security in the midst of so much apparent loss and decay that is the spring of the hope in eternity that is so widely spread among so many of the great religions. To be sure, eternity is not in itself a favourable portent in relation to the question "What may we hope?" It is consistent with despair and wretchedness, or with an acquiescence that may be merely numb, not joyful. On the other hand, without it we may seem to be left with luck and nothing more.

The theological implications of such a theory would be that although God could never be the end (since there is none), he would be liker the later than the earlier stages of the process of becoming, always supposing that his perfection meant simply the fulness of his being. Advancement in any other sense, growth in grace, or in gladness, or in excellence would not be implied by the theory (which, indeed, would be equally consistent with deterioration). Certain philosophers, however, seem to take comfort in the reflection that nothing can be wholly lost, and others discern a ray of joy in the mere reflection that there is an onward process, whether or not it must be upward.

For the most part those who put their trust in eternal being

[1] *Scientific Thought*, p. 82.

Theism and Cosmology

hold that temporal permanence, although distinct from eternity because it is temporal, resembles eternity more closely than anything else that is temporal. "No mark of Time's arrow", it has been said, "is to be found on Aristotle's ultimates."[1] If, they tell us, we choose (fallaciously but very understandably) to employ the language of time about matters that, strictly speaking, are not temporal at all, then it is truer to say that God endures the same for ever and ever than to say that he varies. Hence Spinoza himself, despite his insistence that time is only a form of the imagination, allowed himself the use of temporal language. For instance, he said that the human mind "cannot absolutely perish with the body, but something of it remains that is eternal".[2]

Such statements, if figurative, may have a certain flavour of intelligibility. The unchanging properties in the pattern of renewed existence may seem to be liker eternity than the changing properties in that pattern. As we saw, however, there cannot be more than a certain precarious similitude. The change which is process cannot be generated from anything that is changeless in all respects. That is forbidden by the invincibility of the time-like.

In the scholium to the proposition I have quoted, Spinoza made the emphatic declaration that although we have no memory of our existence before we were born, "nihilominus sentimus, experimurque nos aeternos esse". The reason he gave was that intellect as well as memory transcends the fleeting actuality of sensation. Consequently by the mere fact of transcending sense we ourselves become un-sense-like, and may even become timeless like timeless concepts.

That, I think, is a very common argument, but it seems to me to have no logical force. The man who knows something need not acquire the properties of the thing he knows. If he knows that Caesar crossed the Rubicon he need not become a Roman general with legions at his command. He does not become triangular by knowing a triangle. These statements,

[1] J. L. Stocks, *Reason and Intuition*, p. 64. [2] *Ethics*, V, xxiii.

Eternity

it is true, are so elementary as to seem childish; but they are also very pertinent.

We have all been told that the things that are seen are temporal, but that the things that are unseen are eternal. St. Paul, when he made this famous assertion,[1] may have meant only to contrast the perishable things that could be seen at the time when he spoke with the imperishable things that would be seen, very shortly, when the millennium came. But let us interpret him in the more metaphysical and less eschatological way with which most of us have long been familiar. In that case the philosophy of his assertion would be that sensibles are visible but vanish in the twinkling of an eye, while non-sensible reality is timeless. There would be a procession of vanishing sensibles, but there would also be another world which is not of sense, and does not have the property of vanishing peculiar to sensibles. A further step in the same philosophy would be taken if it were declared that since we are not a mere succession of sensibles, we therefore belong to an eternal over-world.

It is surely a very bold assertion to maintain that nothing except a sensation is transitory at all, or that human beings cannot be transitory because and in so far as they are made up of something more than sense. That would be a very short cut to our unending pre- and post-existence.

But let us pass the point. The doctrine, if true, would be a philosophy not merely of epistemological objects but also of the identity of mental existence with such epistemological objects. If the first of these were true, the second might be persuasive, but since the first is false the second cannot be persuasively sustained. To speak generally, there is no reason at all for denying that we may become acquainted with things that endure by means of sensations that do not endure, or for making a mystery of the circumstance that we may become acquainted with very stubborn continuants through fleeting recollections and intermittent intellectual activities. There

[1] 2 Cor. iv. 18.

Theism and Cosmology

may not be any eternal things; but, if there were, there would be no reason why we should not apprehend them temporally with the transitory attention of transitory beings.

Accordingly the notion of permanence within succession seems always to oust the notion of the timelessness of succession itself. All existence comes to be. Therefore, since its becoming is absolute, the conception of permanence applies determinately to what is present and to what is past at any moment, and applies only conjecturally, at any moment, to what is presumed to be future at that moment. The doctrine of continuants, inevitable from many standpoints, is a doctrine of things with a relatively enduring pattern yet of things that, like everything else, have come to be. It is always an *ex post facto* conception. It does not annul and it should never be allowed to obscure the absoluteness of becoming in the process of actuality.

This conclusion has important theological consequences. It denies determinism in its usual sense, and therefore denies that every event in the world is pre-ordained by divine decree. It denies divine foreknowledge of the future. It denies that God can be Alpha and Omega in one—"at once archetype, intermediary and goal", as *Doctrine in the Church of England* puts the point.[1] These, obviously, are matters of gravity and it is well to be clear about them. They seem to me to be simply obscured in orotund phrases like Dryden's "Till rolling time is lost in round eternity".

According to determinism, everything is objectively immortal at all times. In other words, everything is determinate whether past, present or future, and is so interconnected that everything is fixed, not merely in itself, but also relatively to precedent and to subsequent fact. Inferential prediction and inferential post-diction are metaphysically on the same footing and are possible in the same sense if the data are sufficient. Time may indeed have a "sense"—"Time's Arrow" as Sir Arthur Eddington calls it—but that circumstance does not

[1] p. 79.

Eternity

affect the validity of deterministic inference. Any valid inference of this type fixes the properties of the future as easily and as certainly as it fixes any others.

If, however, the future be indeterminate before it occurs, it cannot be fixed before it occurs. For there is nothing determinate to fix. Hence inferential prediction has quite a different status from inferential post-diction (and also, of course, from memory which is the non-inferential recognition of a determinate past). We might say, indeed, that there would be no practical difference according to this theory if we made the additional assumption that the future would resemble the past. That is the doctrine of practical or experimental determinism, and I am not discussing it now. Metaphysical determinism, on the other hand, declares that the assumption made by practical or experimental determinism is not an assumption but an immitigable truth. I am denying this piece of metaphysics, that is, I am denying its ultimacy. Another way of putting my contention is to say that I have to deny it of God as well as of all else.

Similarly (as I have said) I have to deny the possibility of divine foreknowledge, direct or inferential. Omniscience itself can be applied only to what there is to know, and on the view I am expressing, the sum of what there is to know at any time involves a covert reference to the date of the knowing. It is the sum of what is determinate at any time, and has been determinate up to that time. It cannot, at any moment, literally and directly, include the knowledge of truths about later events before they occur. Foreknowledge, of course, is wider than inferential foreknowledge. If the future at any moment were determinate, God might have direct non-inferential pre-cognition of it in a way that was similar to direct non-inferential retro-cognition in human memory. In that case he would have fore*knowledge* that was not fore*inference*. But if the future is not determinate before it occurs it cannot be foreknown, even by an omniscient being, in *any* correct sense of foreknowledge.

Theism and Cosmology

Again, I have to deny that God can be the beginning and the end in one, if this statement means that his being includes all the future as well as all the present and all the past. A whole, it is true, may have properties that its parts do not possess. The bundle may be strong although the individual sticks are weak. Every whole, however, must be a determinate whole, and the supposed whole consisting of the present, all the past and all the future is not determinate as respects the future. Therefore God cannot be *that* whole; for there never is any such whole. At any moment a pantheistic God might be both the Alpha and the Omega of all that had happened up to and including that moment. That might be true even if he had had no beginning. But it would not be true of the future. We dare not say with Augustine that for God *futura iam facta sunt*.

I may be asked in conclusion whether the views I have been expressing are more than mere speculations. If they were true the theological consequences might be what I say they are, but that hypothetical statement need not be of great consequence if the whole doctrine is just one speculation among a host of other and of very different ones.

About that I would say, firstly (as I said in the last lecture), that the perplexities of this theme are notoriously great, and that I should not like to speak over-confidently. Subject to this very necessary proviso, I should say that I am attempting, with the backing of much recent and of some ancient philosophy, to describe an ultimate property of process as such instead of constructing a theory or an hypothesis about it. If I am wrong, I am wrong, but I should like to present certain considerations which may incline opponents to attend rather closely to the matter, with special reference to the metaphysical status of the future.

In the first place the description I have offered gives a metaphysical or absolute meaning to the fact of *coming-to-be*. It is not clear how a process could be proceeding if it were forever fixed, or be in the making if it were eternally made.

Eternity

If the events at any time are fixed and determinate for all time, time would seem to be eviscerated, and a time without guts would not seem to be time at all. There would be a series of positions called "earlier" and "later", but these would be but names for certain fixed positions. There would be no ultimate freshness in the present moment.

In the second place we have to remember that any description of *coming-to-be* has more than a physical application. It also applies to our own conscious lives. For they too are in process. I do not suggest that our experience of succession is simply or primarily a private matter, projected upon foreign bodies. If it were so projected, and if the foreign bodies were timeless, the projection would be an illusion and we should be left alone each with his own private tempo. That would contradict the very meaning of the alleged projection. Nevertheless, the fact that time-like process is consciously lived through as well as consciously observed does make a relevant difference. For *we* exist and are in process of becoming.

The relevant difference is this. There is no special difficulty in supposing that what seems to be fresh in a foreign body is in fact stale. That would be true of any historian's discovery. It would also be true if the astronomers are right and the sun when we see it is always eight minutes old. But the present, the *now*, in our own experience, even when we seem to be doing the same old thing in the same old way is surely vitally different from all other moments. It is the moment of actuality. It is the moment in which *we* experience ourselves to be in the making. Not to be in the making is to be dead, and we cannot antecede the moment of actuality. I suggest, then, that there is metaphysical death if reality be not also in the making. It can never be completed and live, and the objective immortality of all reality is substantially an obituary conception.

In the third place I should like to point out that if this account of the future be true, a mystery would disappear that without some such clarifying explanation would be wholly inexplicable. Why is it that our acquaintance with the present and with the past is so very different from our acquaintance with the future? In memory we are acquainted with past events quite non-inferentially. We may, it is true, *also* draw retrospective inferences from the present and from the near-past; but memory is independent of these inferences and may be used to verify some of them. To this power of seeing backwards there is no corresponding power of seeing forwards. (There is only prospective surmising.) This statement, I know, would be challenged by prophets, clairvoyants and by such writers as Mr. J. W. Dunne in *An Experiment with Time*. I trust, however, that I shall not be accused of reprehensible dogmatism if I say that one of the most striking features of our acquaintance with temporal succession is the utter absence of non-inferential acquaintance with the future and the manifest presence of non-inferential acquaintance with the past in memory. Our memory, it is true, seems to be restricted to the events in the past that we ourselves have formerly observed; but we are directly acquainted very frequently indeed with our own past; and we can only expect, conjecture or surmise our own future.

If, as I have maintained, coming events are not determinate before they occur, there would *have to be* this difference between precognition and retro-cognition. The difference itself would be a case of *res ipsa loquitur*. We could not foresee the future, because only the determinate can be there to see, and the future would *not* be determinate. All that could be done in this kind would be to make conjectures and to form expectations on the assumption that the future will resemble the past—a reasonable assumption, it may be, since we know, on the evidence of memory, that, in the past, the later did resemble the earlier in very well-marked ways.

Mysteries, no doubt, may occur. If we had no explanation to offer of the plain empirical difference between precognition and retro-cognition, we might without inconsistency fall back upon mere metaphysical wistfulness, lamenting the fact that we are entirely destitute of a power that might have been so very

Eternity

useful. (Or would it? If the future were fixed before it occurred, how could we unfix it in advance?) Humanity would be lucky because its competitors would also have failed to develop the faculty in any marked degree. It is, however, much easier to believe that the absence of the faculty has the best of all possible reasons, namely that, from the nature of the case, there could not be such a faculty, even in a God.

VI

Ubiquity

In the last lecture I spoke of the stubbornness of change, succession or process in metaphysics and therefore in natural theology. I maintained, indeed, that the stubbornness of process was metaphysically invincible, and that deity could neither be exempt from it nor transcend it. We have now to consider whether similar arguments apply to God's omnipresence in space, that is, to his ubiquity.

We should be very unwise, I think, if we allowed ourselves to assume too readily that our findings about time must be matched by similar conclusions regarding space. In the measurement of physical motion, it is true, space and time are closely allied, and, in modern physical theory, are hyphenated into space-time. Psychological time, however, that is to say, the succession in a man's personal conscious experience has several distinctive features that are omitted from the t^1 and t^2 of physical chronometry. These distinctive features might have metaphysical significance even if it were true in fact that there was no mind, divine or human, that was not embodied and therefore spatial. There is, however, a further and a still more important reason for caution about this subject. I have argued that change or process cannot be altogether illusory since even the illusion of change implies the reality of some change somewhere. There is no such knock-down argument in the case of space. Space might be an illusion, and if it were a *phenomenon bene fundatum*, the reality, although it would have a certain correspondence with apparent spatial distinctions, need not itself be spatial. Hence we are not entitled to bracket space and time together for all metaphysical

Ubiquity

and theological purposes, and should try to establish our conclusions separately. The *intellectus cosmotheoreticus*, we may readily grant, tends to accept God's ubiquity in all cosmological theism; but even in cosmology we should not presume God's ubiquity.

In this matter theological thought in the Western tradition has wavered, and wavered again, for reasons that are readily apparent. On the one hand, theologians have been naturally reluctant to deny God's spatial omnipresence or even to appear to do so. If God were not omnipresent there might seem to be something that he did not touch, something Godless that persisted as a going concern at any rate so long as God left it alone. Again, bodies "out there" are to the minds in the street the best if not the only clear examples of what actually exists. Admit that there are such things and that God is not literally in them or they in him, and you may seem to have admitted that there are authentic realities, indeed the very paradigms of the common idea of reality, that are not at all divine. Divine spaciousness, consequently, is a very hard thing to surrender. Most theologians cling to God's immanence as well as to his transcendence.

On the other hand there has been vehement and widespread theological repugnance to the idea that God could in any true sense be material; and opinion seems to have hardened against the possibility of Christian materialism to such an extent that God's immateriality is usually understood to imply the absolute denial that he is spatial at all.

There have, of course, been certain Christian materialists. Tertullian was one, Gassendi was another, Priestley was a third, but the current of later opinion flowed strongly in the opposite direction till in the end Christians tended to assume that it was part of their creed that the very idea of a thinking body implied a contradiction. In the seventeenth century, however, this extreme position seems to have been distinctly novel to the general mind of Europe. Locke may be regarded as a good witness. In his controversy with the Bishop of

Worcester he said: "I know nobody before Des Cartes that ever pretended to show that there was any contradiction in it. So that at the worst my not being able to see in matter any such incapacity ... makes me opposite only to the Cartesians".[1] Locke held in short, that although as it happened, matter in all probability didn't think, and didn't have any other spiritual properties, there was no convincing metaphysical reason why it couldn't. God might have "annexed" such a power to it, or to some of it. Again, there was a still wider disposition to believe that God was in some sense spatial although not in the ordinary sense, and it was very commonly held, as it is quite commonly held to-day in circles outside theology, that nothing is wholly bodiless, and that if there are discarnate spirits which do not have fleshly bodies there are no spirits that do not have at least a "spiritual" body, whatever that may mean.

If "materialism" be interpreted, quite untechnically and quite generally, as the doctrine that whatever exists occupies room, that is to say is extended and is not simply the empty extension of the void, it would not follow in any way that there are any spiritual heights to which "matter" in that general sense may not climb. Consider, for instance, the lot of mankind. It is absurd to say that men do not think, do not aspire, do not admire. Yet men have bodies. It may indeed be a wrong opinion that any of the properties of matter in motion could explain (let us say) why an undistributed middle is a fallacy. None the less there need be no fallacy in the opinion (whether or not it be true) that nothing can detect an undistributed middle unless it is alive, awake and ultimately, a body. The truth may be that when a materialist speaks of his body as being "vigilant" in an intelligent way, what he ought to say is that there is a mind-body partnership, and that the mental partner alone is "vigilant". Yet Locke, too, may have been right when he held that there was no contradiction in the view that the man's *body* might think.

[1] *Works*, tenth edition, IV, 469.

Ubiquity

More generally there may be prejudice without a scintilla of truth in the belief that there is anything inferior, evil, despicable or mean in the mere fact of being or of having a body. If that be mere prejudice, God would not be conceived meanly or despicably if he were conceived as embodied, and it is hard to understand how anyone who believes in divine incarnation could think so.

The sort of "materialism" that theists (and also many atheists) emphatically reject is the doctrine that spiritual properties are inconsistent with material existence, and that "matter" is, as we say, "mere matter". This is fleshly materialism if it is the flesh that is contrasted with the spirit. It is corporeal materialism if the contrast is between body and spirit. Such a theory, however, would seem to be self-destructive. *Homo cogitat.* Therefore if the existence of human thought is a denial of *mere* materialism or simple somatism, mere materialism cannot be a tenable universal metaphysics. Even the illusion of thought would contradict *such* a materialism. A fleshly materialism may indeed be rather more intractable. We may think of the "flesh" rather widely as including all sensual influences, wishes, fancies and programmes. It would be harder to show that all ideas and ideals that are not of this order are empty and vain, but the presumption is that, since there really is a contrast, neither of the contrasted terms exhausts existence. It is clearly fallacious to argue that because certain types of bodies may be incapable of logical or of aesthetic insight, therefore all bodies are so incapable, and the same fallacy would remain if, instead of "bodies", we wrote "the fleshly".

If materialism were a tenable metaphysical theory we could not, of course, reject it simply because we deplored its consequences. We should have to face it along with its consequences. What I have said above is something quite different. It is that a somatism that denies the existence of spiritual events, or a fleshly materialism, a sarcousness, that makes the same denial, is not a tenable metaphysics, and therefore need not trouble

Theism and Cosmology

us. I have further suggested, however, that the mere fact, if fact it be, that *no* spirits, divine or human, are wholly immaterial, but that everything has bodily properties in the literal sense, is not in itself an insuperable barrier either to the spirituality or to the deiformity of reality. There might be minding, aspiring and even divine bodies.

Indeed, it may not be a barrier at all, although, in that case, other and less superficial arguments would have to be met. One of these would be that divine spaciousness is incongruous with divine unity. Space, we may be told, is the theatre of dispersion, separation and estrangement. It may not be so tragic a separator as time, but it permits aloofness; for Scotland is aloof from Antarctica and from Sirius. The reply, however, is clear. However remote Antarctica may be from Scotland, it may be incomparably closer to Scotland than anything that is not even spatially connected with Scotland. Imagine a plurality of entities that were not spatially connected in any way and let one of them be Scotland. In comparison with such estrangement Antarctica might be said to be just round the corner.

That, however, is a fantastic answer to a fictitious difficulty. No cosmologist has the right to wring his hands at the idea of spatial estrangement. His business is to describe and accept the sort of unity and the sort of aloofness that he finds in things. The legitimate demands of cosmic unity must be fulfilled within a spatial cosmos; for the cosmos *is* spatial. Philosophical theologians must be prepared to proceed in the same spirit when they examine the relations between God and the cosmos.

In the first part of this lecture (*a*) I propose to enquire into space and its properties. In the second (*b*), building upon the first, I shall make certain observations concerning divine ubiquity.

(*a*) Let us, then, examine the phenomenology of space.

(1) In the first place we should note that all spaciousness contains features that are perceptually evidenced, and are

Ubiquity

evidenced in no other way. Much in geometry, it is true, may be expressed in numerical equations. Hence some have held that pure geometry can be arithmetized, and its arithmetic eventually logicized; but every space is more than a merely logical structure. The same formal structure might be found in series that were not spatial. For example time, as mathematically conceived, has the same formal structure as a spatial line, but time is an affair of instants while space is an affair of points, and the relevant relations, in the case of time, are the wholly non-spatial relations of earlier and later. The same would be true if we experimented with another method. It has been suggested that a point should be defined as the limit of an inclusion-series. There are, however, many different types of inclusion, for example, the inclusion of logical classes, the over-lapping of durations, and spatial inclusion. Only the last of these would define a point in the sensa that is required, and it is perception (or evidence of the perceptual order as in the case of dreams) that yields the relevant sense. "Space" may be charged and even supercharged with non-perceptual principles and methods of interpretation, but nothing except perception or its like could possibly yield the distinctive spatial *quale* with which we are undoubtedly acquainted. We may contemplate, conceive and symbolize spaciousness, but there is no such thing as a conceptual or symbolic spaciousness immune from all traces of the spaciousness revealed in perception.

(2) In the second place, if we ask what in the way of spaciousness is perceptually revealed, we have to exercise a good deal of circumspection. If by spaciousness we mean extension, spread-out-ness, *partes extra partes*, some of our sensa have spaciousness written upon their faces and others have not. Take the sense of hearing, for example. Certainly we "locate" sounds; but in doing so, do we not usually "locate" the source from which we believe the sounds have come, for the most part employing visual terms? It seems perfectly plain that we do not discern anything that can

intelligibly be called an audible shape or figure. Indeed it would seem that sight, touch and organic sensibility are the only senses that literally and plainly reveal extension. The extended character of the other sensations is either dubious or, at the best, indirect.

Again, there may be a certain nebulousness about the extension that is revealed, say in organic sensibility. Mr. Somerset Maugham has recently poked fun at philosophers for their predilection for arguing about toothache; but let us follow the fashion. An aching gum or the cavity of a tooth that has been recently drawn is ostensibly voluminous and spread out, but its volume is sensed in a much less sophisticated fashion than these phrases sometimes suggest. Perhaps our organic sensations, faithfully reported, do not usually reveal more than a vague magma, massive, unordered and shapeless.

Indeed psychologists, in the days in which such discussions were fashionable, drew an important distinction between what they called mere "extensity", on the one hand, and developed "extension" upon the other. The former, they said, was a vague massy voluminousness, and might be illustrated by the experience of a blind-folded subject when a postage stamp, or a small piece of damp linen, was laid upon one of the more insensitive areas of his skin. It might also be illustrated by the roomy chaos that a man perceives when he is about to faint. The lesson that was drawn from such instances as these was that certain perceptible types of spread-out-ness have very little spatial *order*. They are without definite shape, definite size, definite spatial distance. That is what was meant by mere "extensity." If extension, as opposed to extensity, be regarded as *ordered* spaciousness, the conclusion would be that we often perceive extensity and not extension. Dr. Ward, indeed, maintained that extension is developed in a complicated way from a primitive basis of mere extensity, the mediating influences being what Lotze had called "local signs" and active movements of exploration with tongue and arms and finger-tips.

There are other routes towards a similar terminus. Berkeley,

Ubiquity

as is well known, held that it was absurd to try to join a visual inch to a tactual inch and expect to obtain a line of two inches. Here he exaggerated an important truth. We do conjoin tactual with visual space when, for instance, we explore with the hand inside a cupboard when we cannot see into the cupboard. It is not true, as Berkeley held, that sight-space and touch-space have nothing at all in common. A blind man's geometry corresponds very closely indeed with a seeing man's. In a large measure, however, Berkeley was right. If we keep strictly to what is seen, on the one hand, and to what is touched, on the other hand, we find the greatest possible difficulty in accepting the literal identity of visual and tactual space. There is a marked correspondence, and there are various inferable common relations; but there is nowhere near literal identity.

Again, even if it were conceded as at least a theoretical possibility that each private percipient does in fact perceive portions of a single public world-space whenever he opens his eyes or explores with his fingers, it is clear that the assemblage of these private glimpses and "feels" would constitute not a single space but an untidy mass of spaces. Properly speaking, there would not even be juxtaposition, much less amalgamation. In vision, for instance, we do not simply add or juxtapose two private glimpses when we go out of a house to look at the sky, and although our organic sensibility seems to have a relatively constant horizon within the epidermis, we have already noticed how vague such spaciousness very frequently is.

The inference is that although perception, or a process of the perceptual type, has or may have distinguishing features that mark the difference between spaciousness and any merely logical structure, the "space" of the world is not directly perceptible, but has to be grasped, if it is grasped at all, in a fashion that is largely indirect and is also elaborately inferential.

(3) This leads to a third point. Spaciousness, it may be

Theism and Cosmology

argued, is neither an assemblage of crude perceptual data, nor a bare inference from such data. On the contrary it is ascertained by a unique sort of apprehension that is present in perception although it does not exhaust that process. That, as I understand him, was Samuel Alexander's view. "Every sensory act", Alexander said,[1] "contains in itself, and consequently conceals or masks, a simpler act of intuition." The word "intuition", being a sort of epistemological jack-pot, may provoke misgivings. In a general way, however, we may regard this method of approach as distinct from any we have considered hitherto.

The contention, I believe, is that within perceptual experience there is a special type of insight, having various degrees of perspicacity, into spatial orderliness. It is present in sight, touch and bodily feeling, pervading all the three; but it is never a common residuum that can be distilled out of all the three. Such insight should not in principle be disturbed by the difference between dream and waking, between one special sense and another, between the (so-called) private spaces of different percipients. Moreover it is concrete, not abstract, and is never a naked logical structure. On the other hand, it readily submits to a process of logical *construing*. Hence we are asked to infer that such a rule of construing, whether it be shown in the pattern of some perception, or in freer intellectual exploration, or in conceptual scientific patterns, is all that "space" in the singular should be taken to mean.

(4) A view of this kind may seem disturbingly inadequate to common sense although congruent with much in modern physics. Even the plainest of men, however, might be forced to admit what we now remark in the fourth place, namely that the single "space" of common speech is reached by a process of thinking based upon perception rather than by simple perception, and that some of its implications are of a highly speculative order although we may not usually be aware of the fact.

[1] *Space, Time and Deity*, II, 148.

Ubiquity

That is apparent in the assumptions that underlie the cruder forms of spatial measurement, the chief of which is that it is possible to discover constant spatial units that can be moved about without altering their size or their shape, and can be equated with the measured body. Indeed the ultimate common-sense assumption is that *no* body alters its size or its shape by the mere fact of altering its place. That is why our fathers regarded the Lorentz-Fitzgerald contraction as at the best a fictitious device that could not possibly be an accurate description of fact. In their opinion, any given body occupied a certain definite room at any given time even if it were as soft as butter or as fluid as steam. The room that was so occupied, they believed, was wholly indifferent to the type of body that occupied it, and was inhabited, like a lodging house, by a succession of different tenants. There might be disputes concerning the question whether the room could ever be untenanted, but room, as such, was supposed to exercise no causal influence whatsoever.

This conception, it is plain, owed at least as much to reflection as to sensible appearance. The mere fact that there *seemed* to be roominess might be very unimportant indeed. Indeed the apparent roominess of illusions, dream-spaces and the like was usually erased from the picture. What was left, it was supposed, was "real" space in which a real body had real shape. Indeed the unity of "space" was an implication of the unity of causal influence in physics rather than a given datum to which all physical theories had to conform. It was not a Platonic "receptacle" prior to physics but was excogitated, with varying success, in the course of physical enquiry.

(5) In all this discussion, I have been leading up to an enquiry into "space", by means of an account of the spaciousness that we experience. It seemed to me to be important to do so in order to avoid, if possible, the illusion of premature simplicity. We have now arrived, however, at the conception of "space" as the room in which bodies act, and may legitimately concentrate our attention upon it, even if our confi-

dence is a little shaken by the remarks that have gone before.

The conception of space as the room in which bodies act, when further developed, is closely connected with a familiar metaphysical doctrine, namely that space is absolute, possibly empty in places, homogeneous, continuous and infinite. These characteristics hang together pretty closely although it is logically possible to deny some of them and retain some of the others. If space be the room in which bodies move and act, it is readily conceived as an absolute presupposition of bodily movement and action. If it is equally hospitable to all moving bodies in all directions of movement, it is homogeneous. It is continuous if physical motion be continuous, and it is infinite because any given figure has a surrounding expanse. It might or might not be empty. Since motion in a plenum is possible it might always be filled, containing no unoccupied places. On the other hand a vacuum would be conceivable in any of the three senses in which the vacuum theory has been maintained, namely the vast inane, or pores of vacuity called *vacua disseminata*, or heaped vacuities called *vacua coacervata*. There would be nothing in the nature of mere room to show whether it was occupied or not, although, if space were the theatre of all action, it might be maintained that energy and tension must be ubiquitous.

It is notorious that this account of space has been challenged both on metaphysical and on more empirical scientific grounds for a very long time, and that the revolutionary physics of the present century clamours for its supersession. We may first consider the metaphysical grounds.

These were, in the main, that space so conceived would be a natural anomaly, something real that was also void, brimful of distinctions without any real difference; and, besides, an actual infinite. Itself unsubstantial (or, at least, very unlike a substance since it could not act) it nevertheless was *something* and might even exist *per se*. Being homogeneous, it could contain no determining difference between one part and another part of the vast spread of its *partes extra partes*, and

Ubiquity

some philosophers have held that things that are not discernibly different must plainly be the same. Again it was said to be viciously infinite because it was merely boundless in a negative way without any intelligible unity of self-completion.

It may be doubted whether these metaphysical difficulties were really very serious. There may be much in the universe that is truly *sui generis*. Therefore if space were a substance in some senses of that word (e.g. if it were always subject and never predicate) and were not a substance in some other senses (e.g. if it could neither act nor react) the remedy would seem to lie in a closer analysis of the doctrine that all existence is substantial existence. Again there is no logical contradiction in the existence of things that differ *solo numero* (as newly minted coins appear to do), and it seems rash to declare that there is a metaphysical impossibility where there is none in logic. Moreover, an actual infinite does not appear to be self-contradictory. What is contradictory, as I said in an earlier lecture, is a finite infinite.

The empirical and scientific grounds were, briefly, that "space" of this absolute kind was empirically and scientifically useless. There was no empirical means of distinguishing absolute position in absolute space, not even in Newton's celebrated attempt to prove rotation in absolute space by the experiment of the rotating bucket. For all scientific and also for all practical purposes space had to be treated as a property of bodies. The voice of experience, therefore, was held to applaud the metaphysical theory that space is in bodies, not bodies in space.

On the other hand it is very difficult indeed to believe that there is nothing absolute about spaciousness. If A is spatially distant from B, there must be a spatial interval between them. They are not spatially coincident and they do not touch one another. These statements are absolute even if spaciousness be either a property of bodies or a relation between them. They cannot be upset by any relativity theories, and they refer to *spatial* distance.

Theism and Cosmology

To say, however, that there is something absolute about space is not to say that "absolute space" could exist by itself. If space were a property of bodies there could be no space if there were no bodies. From this it cannot be inferred that if two bodies are spatially distant there must be other bodies between them. There might just be the relation of spatial distance. Again there are several possible views regarding the continuity, the infinity and the homogeneousness of ascertainable spaciousness. The sensible continuity of perceived spaces has always been disputable. If there were *minima visibilia* and *minima tangibilia* with gaps between them too small to be perceived there would be the appearance of continuity without its reality. Relativity theories which pride themselves upon their conformity to the data of observation are not in a position to contest the point. If they demand continuity they must do so for physical reasons that go beyond the data of observation. The scientific question of infinity, again, depends upon the type of spatial relation that seems to conform most closely to the mathematized description of the observed facts. In the Riemannian type of space-relation the curvature is such as to suggest a return upon itself. Hence Riemannian space would be finite. Other theories would suggest spatial infinitude. Once again, the point of view according to which a warped or hummocked geometry determines its own surfaces may be legitimate, but need not be relevant to the distinction between absolute and relative theories of space. It conflicts with Newton's description of absolute space; but that is another story.

Consequently I think we should conclude that although there is something absolute about space, and also something *sui generis* in spatial properties, we are not entitled to infer that there is an absolute space in which all bodies move. It is also quite possible that space is always a property of bodies.

(6) In the sixth place, then, we may ask what we mean by "bodies" when we say that space is a property of bodies, i.e. is either in or between bodies and has no other existence at all.

Ubiquity

By a body we mean a spatial continuant. It has usually a distinctive perceptible contour when it is, as the phrase goes, macroscopic, although that criterion may become dubious when, let us say, ice or wax is melting; and there is no perceptible contour in the case of microphysical bodies. On the other hand, we insist that since all bodies are extended, all the parts of a body must be spatially together. When a body moves it takes all its properties with it. In view of our previous argument regarding temporal process, we have to say that there is absolute becoming of spatial continuants as well as of everything else (if anything else there be). These continuants persist by renewal, and there is no renewal of them before they come to be, that is, before what we call the future. We have also seen that there is a good deal of convention concerning the sense in which any given continuant changes on the one hand, and, on the other hand, gives place to another continuant. That, however, need not matter very much. It is largely an affair of explaining one's conventions. If, however, it were held that all existence is bodily existence we should have to say that everything that occurs is a phase in the existence of some spatial continuant, although many spatial continuants, in passing away, give place to other spatial continuants.

Finite bodies at any given time are always figured. A body, therefore, cannot consist of mere formless stuff which assumes or is invested with this or the other spatial form at this or the other moment. That circumstance constitutes one of the major difficulties in the theory of absolute space. On that theory there is space *and* its filling. If the so-called filling is itself spatial, like a dentist's filling, why do we need absolute space besides? If it is not spatial, how can it be spread out in space? We need not hold with Descartes that space is the essence of bodies. It is one essential attribute; but it need not be the only essential attribute.

The theory of hyphenated space-time in modern physics, commonly called *tout court* the Theory of Relativity, is primarily concerned with the measurement of moving bodies. As such, its interest is technical rather than philosophical. It should be remembered, however, that Newton himself held that geometry "is nothing but that part of universal mechanics which accurately proposes and demonstrates the art of measuring".[1] Like his critics to-day, Newton was dealing not with "pure" geometry but with the general physical geometry of the universe, so far as that was amenable to scientific observation and "deduction from the phenomena". Moreover it seems plain that the "bodies" of materialistic theory, while they may be evidenced by certain privileged sensory observations, are not themselves uninterpreted data of primitive experience. According to Hobbes "the universe, being the aggregate of all bodies, there is no *real* part thereof that is not also *body*; nor anything properly a *body* that is not also part of that aggregate of all *bodies*, the *universe*".[2] We may translate Hobbes rather freely by saying that a "real" body is "real" because it plays its part in the one physical universe. If it is not an executant in that universe it is not a "real" body however corporeal it may look. The mere appearance of extension (in a dream, say) is not evidence of the existence of a "real" body in this sense. Dependable evidence of "real" bodily existence has to be authenticated either by physics or by metaphysics. Even the privileged position of the observations that are accounted reputable in the sciences is governed by some such general theory of the physical universe.

With these lengthy but still abbreviated explanations I shall conclude the first part of this lecture. They were designated as a preamble to the theistic considerations of the second part of the lecture to which I now proceed.

(*b*) At the beginning of our phenomenological preamble we saw that spaciousness was in part ineradicably perceptual, however true it might be both that absolute space is imperceptible and that the space of the world is construed or im-

[1] Preface to the *Principia*.   [2] *Leviathan*, chap. 34.

Ubiquity

puted rather than perceived. The theological implications are plain. If it be held that God's indwelling ubiquity in the world is merely another way of stating what is called the "intelligible extension" of the world, such "intelligible extension" is *less* than what we mean by space. It may characterize but it cannot envelop or encompass the spatial world. For the space of the world cannot be altogether logicized.

Spaciousness, as we saw, is perceptually evidenced in a way that logic and number are not. Its magnitude or so-much-ness is structurally the same as that of other non-spatial series, for example as the magnitude of a temporal series, or of various series of fractions. Consequently there is no such thing as a merely logical space that may be generated by the inner development of a purely logical set of notions. When Mr. Wittgenstein, and some other modern writers, speak of "logical space"[1] they may be using a permissible metaphor, but they are referring to something that is not space at all, namely to the degree of freedom afforded by pure abstract alternatives that apply to everything conceivable and determine nothing that is actual. In comparison "space" is liker colour or sound than the formal logical disjunctive "either-or". It may indeed be logicized or arithmetized to a much greater degree than colour or sound in the present state of our knowledge; but it cannot be logicized or arithmetized without remainder.

Consequently when Sir James Jeans, plunging boldly, as he says, "into the deep waters" suggests that the riddle of cosmology may be solved by the theory that "God for ever geometrizes",[2] such geometry "entering the universe from above instead of from below", and supports his view by the reflection that "the terrestrial pure mathematician does not concern himself with material substance but with pure thought. His creations are not only created by thought but consist of thought", we have to object that geometry is not

[1] *Tractatus Logico-Philosophicus*, I, 13.
[2] *The Mysterious Universe*, chap. v.

pure thought in the sense alleged. This objection, it may be noted, has nothing to do with the philosophical idealism to which Sir James Jeans lends his support. Sense experience may be as much or as little mental as intellectual experience. The objection has to do with the question whether the subject matter of geometry is purely intellectual; and the answer is that it is not.

The same, I think, must be said of one of the most celebrated achievements of Christian philosophy, Malebranche's great argument that we see all things in God who is the place of ideas. According to Malebranche the infinite came first in all our thinking and embraced all that was real. Therefore if we saw the sensible or the material world truly we saw it in God. For it *was* God. He held, it is true, that the sensible world was largely illusion. The senses (he said) taught us to be prudent but did not instruct us regarding the nature of fact. Unlike the senses, arithmetic, algebra and geometry revealed the nature of (created) things. In them we grasped clear ideas, invulnerable in their own kind. Nevertheless, although all that was extended was in God and was seen in him, God himself was not extended. His infinity comprised *intelligible* space, not extension with its divisions, its separations, its aloofness, its *partes extra partes*. The simplicity of the divine nature denied the latter with the same unequivocal firmness as it implied the former. Just as particular numbers, according to Malebranche, were generated by a purely logical process from the *nombres nombrants* of deity, so extension was generated from divine intelligible space.

There is so much logic and so much number in geometry that these ideas make a strong appeal. Spatial types of figure, for instance, seem to have a logical basis very different indeed from the logical basis of certain other natural classes, say the varieties of dogs. Few would suppose that the nature of Dog (or is it the nature of Wolf?) is such as to determine that canine existence must be exhaustively divided into poms, and chows, and poodles and the rest. There have been too many mongrels,

Ubiquity

and the breeders have been far too enterprising for such ideas to be plausible. An exhaustive subdivision of the possible rectilinear figures seems to be far less capricious. It may seem to be generable from the nature of Space itself. If, however, a perceptual element be inexpugnable from spaciousness, geometry is either something less than space or is not, in Malebranche's sense, wholly intelligible. There is also a further notable difference. Any given spatial figure is not merely an instance of spatiality, as being a chow is an instance of being a dog. It is also, as we say, a part of space; and a chow is not a part of Dog in general. Indeed, an "intelligible space" whose parts are not outside one another, seems definitely not to be *a* space at all.

Another part of our phenomenological preamble concerning space dealt with the distinction between the absolute and the relative theories of it. I do not know what the theological implications of the theory of absolute space precisely are. The "receptacle" in Plato's *Timaeus* seems to have been conceived as formless stuff, and may have been in Milton's mind when he thought of the chaos and the night out of which the Creator fashioned "nature" and the earth in the purlieus of heaven. In any case Plato seems to have distinguished space from the "receptacle". Newton, again, seems to have connected "space" with the divine sensorium, but not in a way that lent itself to close theological investigation.

Speaking generally, I suppose that if God could create or annihilate all material bodies, he could also create or annihilate the room in which they operated. Whatever objections there are to the theory would therefore seem to be metaphysical or scientific rather than in any special sense theological. It would seem, however, from our phenomenological preamble that although there is something absolute about spatial relations, absolute space may well be a fiction. When we talk about space in the singular we mean primarily the spatial relations of bodies, and we think of the unity in such relations that seems to be presupposed in what we take to be the single executive order of nature. Such a unity is construed or imputed on the basis of perceptual spaciousness in alliance with its logical structure. We must therefore attempt to construe or impute God's ubiquity upon the same principles—or deny it, if we think it should be denied.

Obviously if, as I am maintaining, there can be no deity unless he is at the least cosmological, God's ubiquity, in some sense, is very hard to deny, whether or not his cosmological influence is to be regarded as a major or as a minor part of his divine activity. Indeed, if by the unity of space we mean primarily the unity in the executive order of what we call the physical world, God, if he operates upon the cosmos, must in some sense be "present" in it spatially. His action may even determine the very order which is what we mean by the unity of space. There might still be a question concerning the manner of his ubiquity, but there could be no question at all about his spatial presence. His spatial presence, on this theory, would mean his dynamical influence, and nothing more. The fact of such influence would be accepted by all theists although they may vigorously debate the manner of it.

Such debates usually concern the manner in which a spiritual being can be said to be "in" a body or to operate "from" a body "upon" a body, the presumption being that deity is either a spiritual or a super-spiritual entity. It has been relatively unusual for philosophers to hold that such a spiritual presence or such spiritual action literally involves the extension of any spiritual being. A few, however, have boldly taken this view, and I think I can discuss the questions I want to discuss in connection with the doctrines of an English philosopher now very little read—I mean Henry More the Cambridge Platonist.

More's opposition to materialism and to Cartesian mechanism was combined with a strong and lifelong belief that all spirits had immaterial extension. That was the principal reason why Hobbes, according to More's biographer,[1] had

[1] R. Ward, *Life of More*, p. 80.

Ubiquity

"been heard to say that if his own philosophy was not true, he knew of none that he should sooner like than More's of Cambridge". Hobbes, in short, agreed entirely with More's attacks upon Cartesian "nullibism", that is, upon the doctrine that anything could exist and yet be nowhere, and Hobbes also thought that More's attempt to correct the inadequacy of materialism, although it seemed to him to be inconclusive, was the reasoned argument of a reasonable man.

The abstract, general possibilities, according to More, were three and three only. God must either be nowhere, or somewhere, or everywhere. Of these possibilities the second, More said, appealed to some delicate minds who did not like to think of God's presence in "miry lakes, dirty ponds and other sordid places", and therefore took upon themselves the ambiguous office of circumscribing God's habitation. Their timidity, More held, was unreasonable, and their scruples about a possible impiety were quite unnecessary. The first abstract possibility was, according to More, plain nonsense, despite the authority of Descartes. Therefore the third of the abstract possibilities could alone be seriously held.

More was utterly contemptuous of Cartesian Nullibism (or Nowhere-ism). "I suspect this new conceit of God's being nowhere", he made one of his interlocutors say in the *Divine Dialogues*, "is the waggish suggestion of some sly and skulking atheists (with which sort of people this present age abounds) who, upon pretence of extolling the Nature of God above the capacity of being so much debased as to be present with any thing that is extended, have thus stretched their wits to the utmost extent to lift the Deity quite out of the universe, they insinuating that which cannot but imply as much in their own judgments. For it is evident that that which is no-where is not at all."[1] Here the interlocutor may have been a little more scurrilous regarding what he took to be Descartes's sly atheism than More would have been had he spoken in his own person; but the principle was wholly More's, and More

[1] *Divine Dialogues*, p. 137.

was tireless in repeating it. He was always convinced that if a spirit were nowhere it could not act upon bodies. As he said, "if the souls of men be nowhere they are as much in one man's body as another's, and one man's soul may move another man's body as well as his own".[1] In short he believed that "the universal opinion of all sober men" was that God is everywhere. Every other opinion was sophistry and illusion.

God's ubiquity, therefore, according to More, had to be interpreted literally. Nevertheless it was a spiritual presence, for God was spirit and immaterial. He had therefore an "immaterial or metaphysical" extension like that of any other spirit. Spirits, More said, possessed a fourth spatial dimension of "essential spissitude," the *binding* dimension of things. This spissitude was supposed to include the "redoubling or contracting of substance into less space than it does sometimes occupy" and More sometimes talked as if that were all he meant by it.[2] If so we might recall Milton's views about the spatiality of angels and of demons, as when Satan "squat like a toad" near the ear of the sleeping Eve assumed quite a different shape when Ithuriel's spear touched him. In other words we should be reminded of a quaint conceit. But More, in the main, had a much more serious contention to offer. According to him all bodies, as defined by the mechanical philosophy, were abstractions from genuine reality. Their motions were not explicable by mere impact. If, *per impossibile*, the universe consisted of such merely mechanical monads, and of these only, it would fly to pieces at once. It would be a stupid and broken congeries. What was needed in addition was a principle of cohesion, a super-mechanical *vinculum substantiale*, a spissitude of invincible integration. That was the "soul" of the universe, the Godhead in it that science and natural philosophy encountered. This soul or Godhead was extended because all that was real was extended, but was "metaphysical

[1] *Divine Dialogues*, p. 146. That, we may say, was his criticism of the Augustinian tag "ubique totus et nusquam locorum".
[2] *Immortality of the Soul*, I, ii, § 11.

and immaterial" because it was commonly omitted from the "mechanical" or "material" definitions of spatial existence. In the preface to his *Antidote against Atheism* More spoke of this spissitude as "vital sympathy and coactivity" and used it in defence both of spiritualism (in the modern sense) and of the *anima mundi*. Regarding the former he was very credulous indeed, but his conception of immaterial extension, although it was very general, was not the product of credulity.

More tried to support his argument by the assertion that even if matter could be thought away or dis-imagined, extension could not. Therefore, he said, "that extension which remains to you whether you will or no, is really and identically coincident with the amplitude of the essence of God".[1] Spiritual extension was the prop and stay of things; mere space, if fallaciously undeified, was even in a sense illusory. Therefore More was not a blind partisan on the side of Absolute Space. "This immense *locus internus*", he said, "the space distinct from matter that we conceive in the mind is a rudish ὑπογραφή, a rather confused and general representation of the essence or essential presence of God, so far as it is distinguished from his life and works."[2] Spiritual amplitude escaped the restrictions of material extension. It was "reduplicative into itself" while material extension was "still single and the same". It had self-unity, self-activity and self-penetrability, and therefore was quite unlike the self-disunity, self-inactivity and self-impenetrability of mere "matter or body".[3] None the less it was genuine extension.

As I have suggested more than once in this lecture, the nullibism of spirit seems quietly to have become a part of orthodox Christian theology, at any rate in the common opinion. Spirits, we are told, are non-spatial, but they certainly exist. They may affect the motions of bodies just as a human spirit, by taking thought, may raise an arm in salute.

---

[1] *Divine Dialogues*, p. 110.
[2] *Enchiridion Metaphysicum*, chap. viii, § 15.
[3] *Divine Dialogues*, pp. 123 ff.

Therefore, it is said, there is no genuine difficulty. A God who is neither in his heaven nor anywhere else, who is crudely misconceived when such language is used even in symbolism, may operate upon the corporeal universe with more than cosmic power, just as a human spirit may move the limbs of "its" body without being either *in* its body or anywhere else.

That I believe has become so firm a tradition that anyone who raises doubts about it tends to be regarded either as a crank or as a crass materialist. Is it not delightfully easy to say that all spatial language about spiritual beings is crudely figurative and so to abandon the affair? To a few of us, however, such statements appear to be slipshod as well as easy, and the abandonment of the problem to be something like metaphysical treason. I also think that the question has not staled with time, and that what More said about the nullibists is as fresh to-day as it ever was. I shall therefore follow his exposition for a little space.

In the *Divine Dialogues* when the Cartesian interlocutor Cuphophron, being hard pressed, became most serious, most metaphysical and most Cartesian, he relied upon two principal arguments. The first was that "the mind of man thinks of such things as are no-where, as of many moral, logical and mathematical truths, which being of the nature to be no-where, the mind that conceives them must necessarily be no-where also".[1] The second was that "cogitation, as cogitation, is *ipso facto* exempted or prescinded from extension . . . forasmuch as we perceive ourselves to think, when we have not the least thought of anything extended".[2]

His critic Hylobares replied (I think with reason) that he "partly understood what Cuphophron would be at but not so fully as to discover any strength at all in his reasonings".[3] Cuphophron's first argument, Hylobares said, would prove as easily that the mind was somewhere when it thought of things that were somewhere as that the mind was nowhere when it thought of things that were nowhere. It was, in short,

[1] p. 151. [2] Ibid. [3] p. 152.

Ubiquity

a complete *non sequitur*. And the second argument was no better. An intensive quantity, such as degree of heat, may belong to an extended body although intensive quantities are not extended, and in any case "it seems a mere sophism to argue from the precision of our thoughts that the things themselves are really prescinded one from another.... When the mind is so set on the metaphysical rack as to pull those things asunder that are found together in nature, and then to say they have no relation to one another ... all conclusions from such principles must be like the principles themselves, defective or distorted."[1]

More, therefore, had very little difficulty in disproving the arguments which purport to show that mind or spirit *must* be unextended, and his rejoinder remains valid to-day. The most that can be shown is the irrelevance of spatial characteristics to sundry true descriptions of *certain* mental or spiritual qualities, and that is no proof of the existence of anything whose properties are wholly non-spatial. In a more general way, nothing is easier and nothing is more futile than the construction of a sham nullibism without either straw or clay. We might ask for instance, "Where is a juggler's dexterity? Is it square or circular? Has it *partes extra partes*?" and might reply that although the juggler's nimble movements are spatial, the property of dexterity is not. That may be a trivial illustration, but it would be no worse than the argument that our minds are out of space altogether because, if they perceive square things, it doesn't follow that they themselves are square.

It is one thing, however, to show that minds *may* be spatial, quite another thing to prove that they *must* be spatial. More's principle here, as we saw, was simply "It is evident that that which is nowhere is not at all".[2] He attempted, it is true, to supplement this argument by a *reductio ad absurdum*, but that is a formal device that can always be employed in cases of alleged self-evidence. In fact, however, this argument of

[1] p. 156. [2] p. 137.

his was only speciously formal and was quite worthless. "To take away all extension", he said "is to reduce a thing only to a mathematical point, which is nothing else but pure negation or non-entity, and there being no medium betwixt extended and non-extended, no more than there is betwixt entity and non-entity, it is plain that if a thing *be* at all, it must be extended." If a thing were literally nowhere it would not occupy so much as a point: but it is begging the question to say that such a thing must be nothing.

Can we then hold that it is evident that "if a thing *be* at all it must be extended"?

In the next century, Hume (who may have been thinking of the argument in the *Divine Dialogues*) pointed out that our sense-perceptions, except in the case of sight and of touch, did not appear to be extended. The materialists therefore sinned against the evidence of their senses when they denied "that an object may exist and yet be nowhere". "I assert", said Hume, "that this is not only possible, but that the greatest part of beings do and must exist after this manner."[1] It was altogether ridiculous to ask whether a sound or a smell was circular or square in its figure, and it was equally ridiculous to ask whether a moral reflection could be placed on the right or the left hand of a passion.

At this point, I think, we should return to the phenomenological preamble that constituted the first part of the present lecture. What, in the end, do we mean by "being somewhere?" If the appeal is to simple perception, as the modern logistical positivists and Hume in this passage have attempted to hold, we should have an untidy mass of perceptions only some of which appear to be extended. If we appeal to "intuition", in Alexander's special sense of that term, we should be appealing, not to simple perception but to an even simpler underlying power which is still a most ambiguous oracle regarding the spaciousness of sounds or of smells. What we do in fact is to use sensory cues in an elaborate process of world-

[1] *Treatise*, Book I, Part IV, § v.

Ubiquity

construing in which the supposed unity of action within the world is the ultimate guiding principle.

Consequently, "to be somewhere" is not a simple statement. It cannot be evident to anyone that "if a thing *be* at all it must be spatial" unless the entire and very sophisticated theory of a single theatre of diacosmic interaction is also evident. When we remember how symbolic, how speculative and sometimes even how nebulous this world-view is at the present time, it would surely be folly to put our trust in the self-evidence of such a principle as More's.

Indeed the abstract notion of existence is far too general to permit of much in the way of self-evidence. I cannot see that we have the right to say "to be is to be somewhere", quite abstractly and generally, any more than we have the right to say, quite abstractly and generally, that to be, *inter alia*, means to be perceived. On the other hand, anyone who says with Berkeley that he cannot attach any meaning to the statement "there is a colour" unless he also means that someone saw it, or to the statement "there is a pain" unless he also means that someone felt it, may be talking very good sense. That seems to me to be the sort of thing that happens in the present case. We are talking about the cosmological aspects of theism. How can we reasonably deny that if divinity operates diacosmically he *is* somewhere, namely wherever he acts?

That is what is asserted by anyone who agrees with More that the metaphysical rack should never be used for pulling asunder what nature declines to divide. Whatever the term "existence" may mean in itself, such an one would say, we *also* mean when we say that God exists that he is efficacious in the world-order. *De facto*, we convey this meaning when we say "there are atoms", "there is life", "there is mind"; and we do so also when we say "there is a God". The point is that all such efficacy implies spatial presence. It is not maintained that everything that in any sense can be called spatial is therefore a "real" existence. Dream-apparitions are spacious but are excluded from the executive order of what we call the world. None the less we construe that executive order on the basis of sense observations. It is not maintained that the precise sense in which the various effective factors in this executive order occupy or permeate "space" is simple to state, or is invariant in all its instances. We find life, for example, in certain material bodies and nowhere else. We cannot say, on that account alone, that life is round or square or shaped like a weasel, but we should not allow ourselves to be duped into the admission that life, for aught we know to the contrary, might exist bodilessly and be nowhere. Certain bodies evince vital properties for a time, and we do not have the faintest reason for supposing that it is quite irrelevant that such bodies are spatial. The same may well be true of the "minding" properties that certain organisms seem to possess.

Let us put the matter in another way. Suppose that the statements "this is alive", "this is mentally alert", do characterize certain organic spatial collectivities in such a way that the executive order of the world is affected by the fact that these animal bodies are neither dead nor drugged. Is there anything substantial to be gained by raising a pother about the question whether life and mind are spread out like paint, or describing some other way in which the behaviour of spatial entities is affected? Such questions might have some importance regarding, let us say, human survival or pre-existence, although, even there, there might be a survival or pre-existence of physical energy which persistently played its part in the world order. Apart from that, the question of spread-outness *versus* some other form of spatiality seems to be profoundly unimportant. Therefore it does not seem to be important for cosmology, which has the executive world-order for its proper province. For the same reason, it should not seriously affect theological cosmology—or "theophysics"—which deals with the divine influence upon or within that executive order.

More's particular theory of a dimension of divine "spissitude" may not have been conspicuously helpful. Its "reduplication into itself" reminds one of the contractility of a sea-anemone, and there would seem to be a logical error in the theory that some particular "dimension" of spatial reality accounts for the integration and cohesion of the whole. Integration or cohesion is not one element among others; and More's conclusion that the body of the universe has a vital and spiritual as well as a merely material constitution may have been hurried and premature in the form in which he stated it. The type of argument is not uncommon. "Matter", it is argued, is defined by its inertia. Therefore motion is "immaterial". So are life and mind. Therefore the negative "immaterial" may be construed positively as a single "dimension" or "substance" or "entity". More's argument is much stronger than this because he revelled in detail, attempting to describe precisely *why* mechanism had to be stuck together with "spissitude". But he may not have been very successful in detail.

On the other hand, he seems to have been wholly justified when he refused to admit that the "immaterial", so conceived, was likely to be non-spatial, and his general belief in the insufficiency of materialism, and especially of the sort of "mechanism" that is based entirely upon interaction *a tergo* and by contact, has many defenders at the present time. One may mention Stout, Smuts, J. S. Haldane and Whitehead among recent writers in English. There is, however, a long way to go from the insufficiency of an explanation in terms of mere inertia to the developed doctrine of an *anima mundi* or to a universal psycho-physical realism.

That is apparent in More's account of the *anima mundi*. "The spirit of nature", he said, "was one and the same everywhere, and acting always alike upon the like occasions, as a clear-minded man, and of a solid judgment, gives always the same verdict in the same circumstances."[1] It may be so, but our assent to such a proposition should not be easily won, and pious men who were also clear-minded and of a solid judg-

[1] *Immortality of the Soul*, III, xiii, § 7.

Theism and Cosmology

ment perceived some of the gaps long before our own time. I quote from an unpublished manuscript of Thomas Reid written for a Glasgow audience towards the end of his long career.[1] "There appears", he said, "to be an immense interval between passive matter on the one hand and active intelligent beings on the other. We can see no such chasm in the works of God. The interval, therefore, may be for what we know (and most probably is) filled up by beings of an intermediate nature which are neither material on the one hand nor sentient and intelligent on the other. And I hope it will appear from what follows in this discourse that however ignorant we may be of the nature and attributes of such beings we have strong reasons to believe their existence." Among the mediators is biological behaviour.

If God be a spirit he need not in consequence be non-spatial, and if he influences the movements and properties of spatial bodies either generally or by the particular providence that is sometimes called a "miracle", he is, in some sense, effectively spatial. That, to repeat the point, is where cosmology and theism touch one another, and the theory of divine ubiquity seems more intelligible than any other. A limited deity, it is true, might have his seat in certain holy places only, and might still be able to operate upon a cosmic scale, much in the way in which, according to certain philosophers, the human mind has its seat in the cortex and optic thalami but affects the entire human body, and, through it, ships and trains and dynamite. There is, however, no sufficient reason for delimiting the zones of direct divine influence in this way, whatever may be true of the physical influence of any human mind. Let us say, then, that God is probably everywhere if he is anywhere, and that if he be nowhere at all, in all relevant senses, cosmology has nothing to do with theology.

If this be a form of pantheism, why should anyone be perturbed? The statement is simply that God is probably everywhere, unless cosmology and theology have nothing to

[1] Now in the possession of Miss Paterson, Birkwood, Banchory.

Ubiquity

do with one another. It would not follow that the marks of deity are equally plain in every corner of the universe, any more than it follows that if something mind-like or something life-like pervades the universe, life and mind may not still be more plainly visible in living plants or in Glasgow citizens than in stones and glaciers. It also does not follow that the universe is alive in the same sense as a cactus or intelligent in the same manner and degree as a Glasgow citizen. The universe might be worse if it were; but it might also be immeasurably better if it were not. There are no simple inferences of this kind to be drawn from the premiss of divine ubiquity; but there is nothing derogatory to the conception of divinity in the idea of a spatial God.

VII

Omnipotence

WE have seen in former lectures that power is essential to deity. There could not be a powerless God, or a God whose power was less than cosmic. Again, although the conception of a limited deity is not absurd, there are obvious reasons why many theologians should want to deny all limitations to the divine power. Hence the conception of omnipotence is readily formed. It is power without limits of any sort, external or internal. It transcends the conception of supreme power, and therefore it seems preferable to many pious minds, Christian as well as Mahommedan, although some may think that a power-theology is more Mahommedan than Christian.

The trouble is that any such abolition of all limitations and conditions—"all these predicates in *non*" as Shelley used to say—may be altogether too sweeping to be intelligible. It may not be more than a magniloquent gesture with no clear positive meaning. On the other hand, the denial of all *external* limitations to divine power does not seem to be open to any such challenge. The meaning would be that there is no power except God's, no foreign rival that could oppose him. Here pantheists would agree with many theists who do not regard themselves as pantheists; and although the conception raises acute moral problems regarding the disobedience of sinners and the existence of suffering which, *ex hypothesi*, omnipotence could avoid, it does not, in itself, raise any cosmological difficulties.

The denial of *all* limitations to God's power, granting that there is no power except his, is (as I have said) not a simple matter, and has been interpreted in a way that would destroy

Omnipotence

the possibility of natural theology. If, for example, it be held, as so great a philosopher as Descartes[1] held, that God could, if he chose, make two and two equal to something different from four, any number of meaningless absurdities could be included in his supposed omnipotence. He could make round squares, vicious saints, tortured creatures who did not suffer, and so forth *in infinitum*.

To this it may be a sufficient answer that a contradiction in terms describes non-entity, and consequently that if God were said to create round squares and vicious saints, he would be said to be creating nothing, i.e. not to be creating. No power is *limited* by the avoidance of self-contradiction.

Such an argument would have a pretty wide range, for it would refer not merely to internal discrepancies in alleged "things" such as a round square, but also to external discrepancies between "things". Thus omnipotence could not decree that the little King of Rome should have been older than Napoleon his father, and, in general, what Leibniz called the compossibility of things would be included in the conception of all genuine power. In view of the extreme difficulty of determining so many of these alleged external discrepancies, the apparent (if ultimately unreal) "limitations" of this order to divine omnipotence might be very numerous indeed. Leibniz illustrated this when he narrated the dream of Theodorus in his *Theodicy*. In the vision, Pallas Athene led Theodorus to the apex of a pyramid where each section revealed one compossible world. Many were fair, but the world at the apex was the fairest of them all; and it was *this* world. Yet it contained Sextus. "If Jupiter", said the goddess, "had here taken a Sextus who was happy at Corinth, or who was King of Thrace, it would no longer have been *this* world. Yet he could not fail to choose this world, which surpasses all

---

[1] Malebranche, who would not have said this about numbers, held that one and the same body might be in two places at the same time (e.g. in Rome and in Paris). If this never happened, the reason was that God decreed otherwise.

Theism and Cosmology

others in perfection, and which forms the apex of the pyramid; otherwise Jupiter would have renounced his wisdom."[1]

Another argument that is sometimes used in this connection is to the effect that omnipotence must be able to create the natures of things, and not merely to create things with a nature. The design of this argument is to try to avoid the sort of consequence that Leibniz drew. There is no pyramid of world-compossibilities, it is said, although there would be if the possibilities were antecedent to God's volitional fiat. Leibniz would have been right if God had had to choose between creating things with such and such a nature, and things with such and such another nature. The truer view is that he creates the nature of each thing that he does create.[2]

This idea may sound attractive. If a human being, for instance, wants a clock to wake him at a certain hour, he has to look for the sort of thing which, arranged in a certain way and following its nature, will act as an alarm-clock acts. Omnipotence, we are told, would not have to pick and choose in this way. It would just create a nature to do the trick. Nevertheless this conception appears to me to be as fantastic as the creation of an integer whose nature was to be both even and odd. Everything that is made must be determinate, that is, must have such and such a nature. God might indeed decree that things with the nature X should be succeeded by things with the nature Y. He might, for instance, decree that the sort of body that moved in accordance with the first law of motion should be succeeded by another sort of body whose motion was circular. In either case, however, he would be making a body with such and such a determinate nature, and he could not override these implications of his own action. He could not, for instance, create a square whose nature was to be round.

We have to hold, then, that no power, not even omnipotence

[1] *Philosophical Writings* (Everyman edition), pp. 257 f.
[2] Cf. Tennant, *Philosophical Theology*, II, 124.

Omnipotence

in any intelligible sense of that word, is, properly speaking, limited or conditioned by being what every existent entity must be, consistent within itself and also consistent with other things in so far as we are entitled to speak of "things" in the plural. This conclusion, however, need not entail precisely the consequences that Leibniz drew, for Leibniz's consequences depended upon the disputable premiss that in any comparison of possible worlds the tiniest difference anywhere must entail an extensive difference everywhere. That premiss might be disputed, and if it were abandoned, very extensive and very fanciful interpretations might be put upon "omnipotence" without any demonstrable disregard of the law of contradiction.

That in itself need not be objectionable. Intellectual play may be very instructive, and although intellectual horse-play is seldom edifying, it need not be altogether reprehensible to stretch one's intellectual muscles and examine what a being with completely unlimited power could conceivably do. If exception be taken to such flights of fancy, what is exceptionable may be the habitual use of this logical playground, and not the occasional use of it. On the other hand, it may well be doubted whether the conception of omnipotence as an agency *capable de tout* is really of serious import in a philosophical enquiry. The notion of omnificence, to borrow a term of Mr. Santayana's, seems to be immensely more serious. If God does all that *is* done, that, surely, is enough. In comparison the problem of what he *could* do, in all possible senses of "could", may reasonably be regarded as a philosophical extravagance. When Christians call God the Father Almighty many of them mean, primarily, that he does all that is done. All power is his power; that is to say, everything that is done, has been done, or will be done, is done by him alone. Others, I admit, would say that God allows much to be done that he himself *doesn't* do, although he *could* do or undo it. So they find a place for sinful actions on the part of mere men. They prefer the notion of potential omnificence to the notion of

Theism and Cosmology

actual omnificence. But omnificence itself would seem to be the central problem.

To be sure, it might be argued that a merely omnificent being might not do very much. There might not be much that was done. Cosmologists, however, are concerned, at the very least, with all that has been done, is done or will be done in what they call the world. That would seem to be quite a lot of action, even admitting that we have no outside standards with which to compare its magnitude. If you are not content with all that has been, is or will be done in the world, what would content you? If you want to show that actual power in its totality is not blind, not feeble, not hesitating, you have the evidence of all cosmological actuality to draw upon. It would be unreasonable (would it not?) to ask for more.

In the discussion that is to follow, therefore, I shall concern myself principally with the notion of omnificence, of all the might that there is; and I shall not further discuss omnipotence in the sense of all that could conceivably be effected. It is disputable whether omnipotence, in this sense, exceeds omnificence when the problem of the meaning of possibility is argued out to its end; but here I shall neglect that question also.

The usual way of attempting to prove divine omnificence is to try to show that all natural causes are "second" causes, and that second causes are not authentic causes but are called "causes" by courtesy only. Therefore if there be powers or causes at all, there must be "first" causes and these must be the seat of all genuine power. From this, it is true, the existence of a *single* First Cause, Omnificence itself, could not be validly inferred. The contrary, indeed, would be true, if, as some theologians hold to-day, and as Plato held in his *Laws* (the tenth Book) every soul were a first cause. On the other hand, the balance of the evidence might strongly incline towards the unity of first causes in an integrated omnificent being, and such omnificence would be divine if it had the other properties that we ascribe to Godhead.

Omnipotence

I am stating these arguments in their most ambitious form. If they are stated less audaciously they are also feebler, and in any case would not prove omnificence although they might support the belief in divine supremacy. I am anxious, in other words, to avoid the form of these arguments in which the deism or the theism in them is little more than official. Thus it may be held that second causes are quite genuine causes but are, in some obscure sense, derivative. A collocation of particles, given the initial divine *chiquenaude*, might continue indefinitely on this borrowed momentum. If, in accordance with the law of entropy, it eventually became stagnant as regards all possible work, the battery might again be recharged by divinity and the world would again become a going concern. Similarly the doctrine of created creators, with an inferior but genuine creative efficiency, contradicts the strict notion of divine omnificence. It has been common in our philosophy since Plato's *Timaeus* and the Arabian Neo-Platonic *Liber de Causis* in the ninth century, but indeed is an old and a far-flung idea in a host of cosmogonies.

The same should be said of the doctrine that second causes are quite genuine but require divine concurrence. In the seventeenth century that was the prime contention of the Molinists (Jesuits). The theism of this view, it is true, was much more than merely official. It resembled what in a different sphere is nowadays called Constitutional Medicine, a revival, or survival, of the doctrine of the *vis medicatrix naturae*. According to this theory, all that medicine or surgery can do is to assist nature. The main fact is a viable constitution. Similarly it might be argued that the main fact about reality was its divinity, or, in other words, God's ultimate power shown in his "concurrence" in nature. Such concurrence, however, would be very different indeed from omnificence, just as the *vis medicatrix naturae* is greatly assisted by the surgeon's art, or by insulin, or by prontosil, and the doctrine would tend to become very official indeed if the analogy were exploited and if it were maintained that the *vis medicatrix naturae* is only a name for the unoccupied places in the advance of medicine and of surgery.

Let us suppose, then, that the important metaphysical question is whether first causes are the only true causes. In that case, it seems necessary to explore the nature of causality and so to distinguish between the efficacy of first causes and the impotence of second causes. That, of course, is an ancient and also a perennial philosophical problem. I shall try to alleviate some of its perplexities by explaining, in advance, that when I speak of causes I include what might more appropriately be called cause-factors. Most events are highly complex. Their entire causes may embrace a large slice of the universe. In practical affairs, when we speak of *the* cause of this or of that, we seldom mean the entire cause, and we often cut our knots instead of untying them. Take for example some legal claim for damages. A shipowner, let us say, claims damages because some other boat, negligently handled, has injured his vessel. Let us also suppose that he claims specially heavy damages, because, in the actual circumstances, he has had to pay at exorbitant rates for chartering a substitute vessel. No one supposes that seamanly negligence was the sole cause of all the events in respect of which such a claim is made, or that the claim for the repair of the damaged boat is on the same footing as the claim for the temporary substitute. Yet claims of this type are made in the courts.

This being understood, there would seem, in the main, to be three ways of interpreting causality, namely invariance in antecedence, activity, and some sort of logical or quasi-logical determination. These three senses need not always be wholly distinct. Indeed each of them may make large drafts upon the credit of one at least of the others. They are, however, distinguishable, and I shall try, in the first instance, to consider each of them separately. Having done so I shall try to find out how far each of them could tolerate or should welcome the conception of a divine first cause.

The invariance theory of causality is to the general effect

Omnipotence

that if B in the accredited past (i.e. in observations that are accounted reputable) has unvaryingly been preceded by A, we are entitled to infer that there is a rule of *invariable* sequence for all time, and that such invariable sequence is all that is meant by causality. Here the inference from *de facto* uniform sequence in the past to invariable sequence for all time may be challenged. The reply would be that if our reasons for believing in causality are dubious, they are, at any rate, the best that there are. Again it seems necessary to stipulate a certain connexiveness in A and B. We do not commonly infer that because the inbreathing of Andrew MacConnachie in Glasgow has always been followed by the outbreathing of a great many Chinamen and Arabs, therefore Mr. MacConnachie has any connection at all with these Chinamen and these Arabs. *Post hoc*, in short, has to be pegged down in various ways before it can even appear to be *propter hoc*, and the peg is usually called "substance" e.g. Mr. MacConnachie's lungs. The pegging down of causality, however, to the transactions of substances in connected sequence is a requirement of the invariance theory and not an objection to it.

The activity view of causality is frequently said to be evidenced in our personal experience and to be extended by a legitimate process of generalization to foreign existents. According to this theory, we know what it is to exert ourselves, and so to influence our own subsequent lives and the history of other things. Here there is sequence, and the sort of pegging down that is essential to the invariance theory, but there is no need to assume *invariable* sequence. On the contrary, we may exert our powers quite capriciously (or so it is said), and we could experience power on a unique occasion. No doubt there are difficulties. If we begin to ask why we should exert ourselves vigorously on some occasions and indolently on others, we might be inclined to reply that a big effort is needed for the first, and that a small effort suffices for the second. If so the conception would be one of the amount of effort just sufficient to bring about a given effect. That, if it were generalized, would resemble the uniformitarian view. An activist, however, need not dig that particular pit for himself, but may cling to his alleged experience of an activity that might be capricious.

The theory that causality is a species of logical tie was so firmly fixed in men's minds by philosophy that it took a great deal of philosophy to unsettle it. To know the causes of things was to know their reasons; and reasons, of course, were supposed to be logical. Hence Spinoza, like almost all the eminent and almost all the negligible philosophers in a great tradition, spoke of "ratio seu causa".[1] Even before his time, however, the seeds of doubt had been sprouting, although not perhaps very visibly, in places not usually sceptical. In the next century Hume, whose scepticism was nourished by a thin residue of rationalistic ideals, succeeded, although very slowly, in radically changing the received opinions upon this matter.

Examiners in philosophy are fond of asking what precisely Hume had proved. With regard to that question I would rather be examiner than candidate. It is clear, however, that Hume convinced many later philosophers that the unvarying sequence of accredited observations in the past was not an adequate ground for inferring the invariable necessity of such a sequence for all time. He also convinced them that there was no logical contradiction in denying that every event must have a cause in the uniformitarian sense, and that there could never be a valid rational intuition independent of repeated observations and clearly showing that A must always produce B rather than C or D or nothing at all. Hume therefore effectively denied that the causal tie was directly or obviously logical; and he maintained that the activity view of causality was a popular superstition whose evidence, such as it was, was merely uniformitarian.

These, I submit, are the principal theories of causality. I propose, therefore, to ask how the distinction between first

[1] *Ethics*, IV, Preface.

Omnipotence

and second causes should be interpreted in terms of each of them in turn.

(1) On the uniformitarian view it seems clear that there is no such distinction among invariable sequences. Any uniformity of sequence, pegged down in the way mentioned above, is as good as any other. It is a uniformity of sequence, and that is all that is meant by the causal relation. Some such uniformities may be broader than others, and some may be firmer than others if they have a wider range of uncontradicted experience to back them. In principle, however, uniformity of sequence is the last word in this matter, and it is just itself.

There are here two points of critical importance. The first is that *invariable* sequence has to be inferred, rashly or not so very rashly, from *unvarying* sequence in past observations. It follows that any sequence admitted to be unique and unparalleled can have no such basis. This conclusion, it is true, must be interpreted with proper caution. Hume himself admitted that a single experiment might be enough to establish a natural law.[1] He meant, in effect, that if the alternatives tested in some particular experiment could be subsumed under some general causal law presumed to be established, the elimination of all but one of them in some novel experiment would destroy the invariance of all except that one. In principle, however, there could be no evidence for the distinction between *post* and *propter hoc* unless *several* uncontradicted instances pointed towards the latter.

The point has obvious importance regarding the supposed origin of the world, which, *ex hypothesi*, is an event wholly unique and unparalleled. It implies, indeed, that a first cause of the world, in the uniformitarian sense of cause, could never be established. On the other hand, the uniformitarian view of causality would not necessarily refuse houseroom to all the senses in which it has been held that there are first causes within the world. If by a first cause were meant simply an uncaused cause, then it would be possible for experience to

[1] *Treatise*, Book I, Part III, § 8.

Theism and Cosmology

reveal entities which although they themselves had no invariable antecedents were nevertheless always followed by unvarying consequents. Such entities would just pop up, but, having popped up, would have uniform consequents. Human selves or human volitions might be instances.

This leads to the second important point. It would be consistent on a theory like Hume's to hold that several uniformities of sequence had been established (although never perhaps *demonstrated*) in nature, that their number, very likely, could be increased, but that the residue of nature might be simply non-causal or chaotic. For the most part, however, uniformitarians (including Hume himself) preferred the opinion that since so many natural events had been observed to be instances of uniform sequence, and since none was known *not* to be an instance of such a sequence, the sensible conclusion was that all natural events are of this type. Every natural event on this view must be presumed to be an instance of causal sequence, that is to say to have uniform successors and also uniform antecedents. It would be an effect in the uniformitarian sense and also a cause in the uniformitarian sense. If so there would be no first causes in nature, no uncaused causes.

(2) If the activity theory of causality makes the generalization, "Everything that has a beginning must have an active cause," its station, in the end, would be similar to the station of the uniformitarian theory in the respect last mentioned. Nothing that had a beginning could be without an active cause. So wide a generalization, however, is not logically required of the activity theory. Its favourite field is volition, and many of its exponents have regarded volition as the absolute beginning of motion or of some other change in body or in mind. As Locke said in his obscure but celebrated chapter on "The idea of power", "the idea of the beginning of motion we have only from reflection on what passes in ourselves where we find by experience that barely by willing it, barely by a thought of the mind, we can move the parts

of our bodies which were before at rest".[1] In his usual cautious but pleasantly inconsistent way, it is true, Locke combined this assertion with the further statement that "whatever change is observed the mind must collect a power somewhere able to make that change," and he allowed "a very obscure idea of an active power" in billiard balls which could "transfer but not produce any motion".[2] But Berkeley "bantered"[3] the second admission, and Hume pilloried the first.

Activists, therefore, very commonly hold that an active cause produces and does not merely transfer its efficacy. It makes a fresh start, and is an author and a sort of artist, not a mere transmitter or a mere resonator. If such activity were attributed to all natural agents, the conclusion would be that *all* causes are first causes, in so far as agency and causality mean the same thing. The logic of the theory, however, would permit a distinction to be drawn between true or originating "first" causes, on the one hand, and, on the other hand, the connexive sequences, sometimes called "causes", that merely depend upon such origins. These would be echoing sequences and so would be opposed to originating agency.

If the views I have expressed in previous lectures regarding the nature of temporal process are sound, there could be no ultimate distinction of this order. My contention was that the moment of actuality always inherits and also creates. According to the activists "creation" is identical with causality, activity or originating power, "inheritance" is passive transmission or echoing. Supposing, however, that such a distinction could effectively be drawn, that there was an ultimate distinction between the origination and the mere persistence of a state of change, between activity and passivity, between first or originating causes and second or echoing quasi-causes, the application of the theory to volition would instil many doubts. If anything in the world seems to be originated, human volitions seem to be originated. To all appearance, they grow out of a man's character. What would have to be

[1] *Essay*, II, xxi, § 4.   [2] *Ibid.*   [3] *Principles*, § 25.

Theism and Cosmology

said, therefore, would be that in so far as they are originated they are not genuinely active, but that nevertheless they are not "a hundred per cent passive" and consequently that they are true originators in the active percentage of their composition. In that case the contention would be that every volition, in some fraction of its being, is a genuinely "first" cause. Its effects on the other hand might never cease although they would usually become diffused and submerged.

(3) According to the theory that a cause is a logical tie, the distinction between first and second causes might seem to be very simple. It would be the same distinction as the distinction between first and second principles, between axiomata prima and axiomata media. Cause and ground would mean the same thing; and ultimate grounds can be distinguished from derivative grounds. The finality of the distinction might sometimes be challenged, but its meaning would be plain enough. We do speak of first principles from which other principles may be deduced. Such deductions might be drawn from mere assumptions, but first principles, according to an important tradition in philosophy, may themselves stand to reason. If so, they would be established by the intuitive reason and their consequences would be drawn out by the illative reason, these two functions being functions of one and the same "reason". In this sense, "first" principles would be underived because they would not be deduced from any other principle. But all other principles might be derivative.

It is plain, however, that even if the relation between cause and effect were a species of the general relation between ground and consequent, it would be a very special sort of species applying only to *sequences* and to these only if they were pegged down in a certain way. The question would therefore be whether first *causes* could reasonably be believed to have the same status as first *principles*, and the answer is that there are very strong reasons for holding that they could not. Let us suppose that there is a logical nexus between the dropping of an incendiary bomb in Barcelona and a confla-

Omnipotence

gration in that city. Such a cause, if it were analogous to a first principle, self-justifying and underived, would have itself to have no relevant antecedents. As we have seen, there are very strong reasons for denying that there are any first causes of that kind in nature; and the idea of a cause *of* nature is not supported by any analysis of the causes *in* nature. The general conception of a logical "first ground", accordingly, would seem to be inapplicable to the special type of logical tie that causality must be if its nexus *is* logical.

In terms of this discussion of the distinction between first and second causes in each of the three principal meanings of causality, it should be possible to consider with some precision whether there are good or overwhelming reasons for believing in the omnificence of a unitary first cause.

As we saw in the last lecture, the unity of the cosmos, physically regarded, means primarily the unity of its power. The so-called unity of space is the unity of the executive order of the world. Mere spaciousness does not suffice of itself, for dreams are spacious in the look of them and yet, taken at their face-value, are not regarded as parts of the world. The dreamer is such a part, and he does dream; but there are no dream-things in the world. Let us assume, however, at least for the purposes of the present argument, that the unity of natural power is not in dispute and enquire into its relations to omnificence.

If causality were simply a name for invariance of sequence, all our evidence regarding it would be obtained from the observation of natural sequences, and these would exhaust our evidence of power. There would be no need to assume that the world itself was a term in such a sequence, and no evidence in favour of that idea. It would be impossible to go behind the fact of such invariant sequence; for that would be the last word in the matter. Within the world it would be reasonable to hold that all that happens exhibits such sequence (although we do not know for certain that this must be the truth) and that there is such a sequence, both *a parte ante*

Theism and Cosmology

and *a parte post*, in all that happens. In other words it would be consistent and not unreasonable to believe that all causes are of the same order, and that the world is patterned in this causal way without beginning or end. We could speak of omnificence if we chose to do so, our meaning being that all the causal strands in nature are part of a grand causal pattern in which there are no gaps and no vagrant threads. Such an "omnificence" would be divine if the grand causal pattern had divine properties. There would, however, be no need to speak of omnificence in any other sense, or to argue to the existence of a non-natural First Cause.

Accordingly, I shall not further consider that particular conception of causality, but shall turn to the others.

In the past, a very high proportion of the philosophical arguments in support of divine omnificence has depended upon an activist theory of power. "Matter", it is said, being defined by its inertia, is only passive. Therefore it cannot be an active cause. Therefore it is impotent. Therefore materialism is a metaphysics of impotence, and consequently an absurdity. According to Locke, in the passage already quoted, the billiard ball "obeys" the billiard cue, but with a wholly passive obedience. The Christian philosophers Malebranche and Berkeley could not accept such a compromise (if it were a compromise) and neither could Hume[1] (who was not a Christian philosopher). The supposed transference of passive power (these authors said) was never observed or legitimately inferred, and Locke's critics added that if it were "collected" from the facts, it was collected by a muddled mind. Indeed Malebranche and Berkeley believed that they could give a very simple and striking proof of divine omnificence. According to Berkeley the transferred or derivative causality attributed to second causes was not, as Locke and so many others had said, "a very obscure idea of *active* power". It was not power at all. Here Malebranche, writing before Locke had written, anticipated Berkeley, as Malebranche himself had been antici-

[1] *Treatise*, Book I, Part III, Sect. xiv.

pated by De La Forge and Cordemoy. There were no second causes to limit divine omnificence, and all the theological talk about God's "concurrence" was based upon a mistake.

Malebranche and Hume further maintained that what was true of the billiard ball was also true of the billiard cue, and of the arm that manipulated it. There also nothing could be observed except succession. Even in ourselves we did not experience the efficacious origination of motion. We observed succession, and the same would be true of the origination of our thought-processes. Here the argument took a turn that Berkeley found very unwelcome; but Malebranche joyfully inferred God's omnificence from the impotence of nature, physical, psycho-physiological and psychical. Hume's conclusion was quite different. It was that there was no experience or other knowledge of efficacy except the experience of constant conjunction in succession. Therefore the attribution of such efficacy to deity was wholly out of the question.

That is one of the major difficulties of a proof of omnificence that is drawn from arguments of this type. The implied conception of efficacy as well as its supposed opposition to the non-efficacy of reputed empirical causes may turn out to be an illusion. Another difficulty is scarcely less formidable. If the notion of efficacy be retained despite these objections there would appear to be far too much efficacy for a comfortable acquiescence in the belief that God has a monopoly of efficacy. If it be granted that human beings do perceive genuine efficacy in their own volitions, then human beings *are* efficacious, although, it may be, on a very small scale. Therefore if human volitions are human and not divine, God is not the only efficacious being, that is to say, he is not omnificent. He might be omnipotent, in the sense that he *could* have avoided creating these efficacious beings; but he could not be omnificent.

It might be replied that although our volitions are plainly efficacious, it is not we who will them but God who willeth them in us. There are many difficulties in that conception,

Theism and Cosmology

but one may suffice for the purposes of our present discussion. The very meaning of the supposed efficacy in our voluntary actions is that *we* experience the relevance of our volitions to the willed events that succeed them. *We* resolve to sign a cheque, and the cheque is signed. If it is not we that sign the cheque but deity instead, how could we know that our volitions have anything to do with the result? If we rely upon the method of concomitant variations our argument is indirect, and is either frankly uniformitarian, or else depends on the elimination of every imaginable causal nexus except the volitional nexus described as God's willing in us. The argument relies on *our* volitions and the supposed revelation of personal efficacy in volition would appear to be inconsistent with the idea that it is God, and not a man himself, who wills.

Accordingly, I do not think that divine omnificence would be a reasonable conclusion if the premiss be that *we* are obviously the authors of our voluntary actions. On the premiss the conclusion could be true only if pantheism were true, and although the premiss may not be wholly inconsistent with every form of pantheism, its plain meaning is that *we*, as finite entities, are obviously the causes of our voluntary actions without any *arrière pensée* regarding our possession of a transfigured and divine dimension in our being. In Western theology, however, it is very frequently maintained that, in the last analysis, every cause must be a volition, and that God must be the cause of everything in the world except the actions of undivine voluntary agents. If men, in certain of their doings, were the only such agents, God, although not strictly omnificent, would be very nearly so, at any rate if human voluntary action is as puny as, in a cosmological sense, it appears to be.

This argument would involve three governing premisses, namely, (1) that everything that happens in the world has an efficacious cause, (2) that our volitions are the sole cause of the actions that we call voluntary, and (3) that there can be no cause except a volition.

Omnipotence

(1) I need not here say much about the first premiss, since I am at present discussing it, directly or indirectly, most of the time. The premiss is denied if it is held that there are any uncaused causes (i.e. first causes in one sense of that term) *within* the universe, and it would not support the idea of a first cause *of* the universe. It is denied by all libertarians, and it is not a necessity of reason. These cursory remarks may here suffice.

(2) The second premiss surely contains a good deal of effrontery. Let us allow, for argument's sake, that we are acquainted, by first-hand experience, with the efficacy of our own volitions, and let us also allow, for argument's sake, that what each man among us finds to be true in his own case may legitimately be believed to be true in the case of every other man. How, on these assumptions, is it at all reasonable to infer that our volitions are the *sole* causes of the ensuing actions that we call voluntary? By a voluntary action we mean, I suppose, one that we can perform if we set ourselves to perform it. That is the usual meaning of the term, and it has an important empirical sense. It distinguishes, for instance, between obedience, on the one hand, and, on the other hand, the sort of behaviour that can neither be obedient nor disobedient since our volitions, as in the case of a man falling from a tower, do not affect the result. But how can we claim that we know for certain what will be efficacious if we set ourselves to do it? And how do we know that, if we do set ourselves to do it, the volition was the sole operative factor? In sudden unexpected paralysis, to choose the stock illustration, a man sets himself to move his arm and the arm does not budge. In this case the volition is not successful; but it is a perfectly genuine volition. We therefore do not know in advance what volitions will be efficacious. We should not know it even if the uniformitarians are wrong in maintaining that our empirical beliefs about what we can and what we cannot do if we will are wholly dependent upon the winnowing lessons of past experience. Suppose, however, (as in the normal

case) that the man is not paralysed and that his arm does move when he decides to move it. Are there not, in all probability, a host of factors, many of them quite unknown, in the case of that simple movement? Have we the least reason for inferring that the volition in such instances is more than one of the cause-factors, among many, that pertain to the initiation of the result?

(3) The answer to the third question goes along with the answer to the second. If a man's volition, in the normal case, is (so far as he can see) only one of the cause-factors in the result that he intends to bring about, and if the other cause factors are *not* his volitions (so far as he can see) what reason can he have for inferring that the other cause-factors, even in the man's own voluntary action, are *anyone's* volitions or are instances of some impersonal volition (if that be possible)? If a man, as the saying goes, "builds better than he knows", is there no efficacy except in the inferior sort of building that he consciously intended? If, intending X, he performs Y, is he then altogether inefficacious? That, surely, would be a very odd doctrine, and although, in the case of a wiser result than was intended, there may be some reason for suspecting the guiding hand of a providence, there is no such reason when the intention is wise and the execution feeble. I need not give illustrations.

As we have seen, a favourite philosophical view is that volition gives us positive experience of causal efficacy that is not mere sequence, and that, once we are in possession of this positive evidence, we are entitled to generalize it throughout nature in distinguishing *propter hoc* from what is merely *post hoc*. The uniformitarians deny that we have this positive revelation, and some other philosophers say that the revelation is so closely interknit with the peculiar experience of volition that no generalization beyond volition (to the inanimate, say) is permissible. There may be virtue in both these houses; but in so far as the alleged efficacy of the volition is supposed to be proved by the success of *that* cause-factor, it is at least permissible to point out that non-voluntary antecedents are also "successful" and that there is no contradiction in the idea. Non-voluntary digestion may be just as successful as voluntary swallowing. If so it is entirely legitimate to argue that although successful volitions reveal the efficacy of a voluntary cause-factor, their success is due in part to other non-voluntary cause-factors, and that, in general, most "successful" cause-factors need not be supposed to be volitional at all.

So in theology. The principal reasons why an activistic (and, in particular, a volitional) theory of causality has been endorsed by so many theologians are two. The first is that a volition is a *mental* cause of effects that need not be mental, and that it is obviously most important for theologians to maintain that God is mind or at least is not below mind. The second is the familiar doctrine that volition and it alone is an uncaused cause, an originator that is not itself originated.

The first contention would be true if volitions are effective cause-factors; and I believe that they are. There is, however, no need for assuming that the spiritual or mental influence either of a man or of his God is limited to voluntary influence. Much of a man's spiritual influence may be quite unintended. It radiates from him, as we say, or, again, it corrupts. The former statement might also be true of deity.

The second contention is an old friend, but, unfortunately, not a trustworthy friend. Let us repeat our objection. Our volitions seem to grow out of our characters, both in their vacillations and in their resolute consistency. There is, indeed, a sense of will—"the stable will," as it is called—that seems to describe the general bent and tendency of a man's character and *not* any *fiat* or express voluntary decision. Thus many philosophers say that the most important things that we do "with a mind" and "with a will" seem inevitable to us and not to be a matter of "making up one's mind". They resemble a conversion that is not sudden in religious affairs: and even a sudden "conversion" may seem rather to happen than to be willed. If it is "willed," the process of willing may even seem to be superficial. I know that there are other points of view; but the whole idea that a volition is obviously and undeniably a "first" cause seems to me to be very seriously suspect.

Let us pass, in the third place, to the arguments that profess to show that there must be a divine omnificent first cause, it being assumed that causation is a logical tie.

On this view, causation is the *logic* of change, and logical relations, it is to be presumed, have to be ascertained, not by sense-observation, but by some intellectual process. Consequently the theory would not be in the least perturbed if the causal tie cannot be seen like a colour or heard like a sound, or observed introspectively as the specific experience of trying or of striving may perhaps be observed.

The question at once arises what this intellectual process is supposed to be. Hume's argument was, in effect, that if the causal tie cannot be observed in sense-perception or in introspection, and is never contained in the analysis of any event, there can be no valid grounds for believing that there is *any* causal logic in change. "The repetition of like objects in like relations of succession and contiguity", he said, "*discovers* nothing new in any of them" for "'twill readily be allowed that the several instances we have of the conjunction of resembling causes and effects are in themselves entirely independent. . . . They are entirely divided by time and place." [1] Hume therefore inferred that the objects did not differ when repeated. It was only *we* who differed by forming expectations, after the experience of repetition, that we did not form before that experience.

This argument, however, may have laid itself open to a flanking movement of an intellectual kind. Let us grant that the causal tie cannot be sensibly perceived. Let us further grant that the mere intellectual analysis of A will never show that it must be succeeded by B, or, indeed, by anything at all. (That would be false, in the second part of it, if every event passed *on* as well as *away*, but let us neglect the point here.)

[1] *Treatise*, Book I, Part III, Section xiv.

Omnipotence

Let us grant, thirdly, that the repetition of like sequences without any contradictory instances, is, *ex hypothesi*, the repetition of precisely similar sequences which, being precisely the same, cannot also differ in any relevant way. All that follows is that we cannot discover anything new *in* the instances so repeated. It does not follow that the repetition does not tell us anything new *about* them. Suppose the contention were that there is a logic in temporal succession, which must be pegged down to specific causal sequences, but that no specific causal nexus could be sensibly perceived or obtained by mere analysis of any event, say A. In that case what the repetition (uncontradicted, and in varying circumstances) might do would be to limit the *a priori* possibilities. The nexus, by hypothesis, is such that it must be the same wherever it occurs. If B succeeds A in varying circumstances, and has never been observed to occur except when A preceded it, the *a priori* possibilities would be very considerably narrowed, although it would still be possible that contrary instances had escaped our notice or might turn up to confound us in the future. Such an argument would be intellectual and indirect. It would also, as I have said, be *about* A and B, but only *about* them. Nevertheless it might be highly relevant.

I am not suggesting that such an argument would be an adequate answer to Hume. It is an argument that presupposes the ubiquity of the causal tie throughout nature, and the logical character of that tie. Both of these presuppositions might be denied, and Hume, in fact, denied them both. I am only saying, hypothetically, that if it were true that there is a causal logic in all natural events, the mere fact that we could not discover such a tie by simple analysis of any given event considered solely in itself would not prove that the indirect methods of presence and absence could teach us nothing at all. The general opinion of logicians at the present time is that methods of elimination are much less decisive, if the universality of the causal principle is assumed, than most philosophers have, in the past, supposed them to be. Elimination would tell us very little unless the number of alternatives was manageably small. The principle of the uniformity of nature does not say anything about the number of natural alternatives, and the assumption of a *simple* "ground-plan of nature" would seem to be either a guess or a stipulation.

Our present question, however, is whether the logical view of causality, if it could be sustained, would supply evidence of divine omnificence. Here the only important theological question would seem to be the argument to a logical First Cause determining all specific logical sub-causes. On that point our previous discussion in the present lecture is directly and obviously relevant. If causal connection in detail be the logic of temporal process in detail our principle would be that any given event N is the logico-temporal sequent of M in particular (and not of K or L) and is the logico-temporal antecedent of some particular event O (and not of W or Z). There would be a specific causal tie in both temporal directions, and there might also be coexistent spreading logical ties, but there would be no place for a Grand First Nexus, a Primordial Universal Tie. The world would be a tissue of specific causal ties. The search for cause-factors would always be a search for threads of continuity within the world-fabric. It would be unconscionable to look for a thread of all threads to constitute the fabric itself.

A reply, however, might be made. After all, it might be said, a cause, if it be the logical tie in temporal sequence, must be the logical ground of its effect. It really must be the reason *for* the effect. Let it be granted that a particular finite cause is the reason for a particular finite effect, and is itself the particular finite effect of an antecedent particular finite cause. In that case, unless there is a whole succession of universes, each logically generating its successor, the cause of the universe could not be a particular finite cause in the sense supposed; and a First Cause of the universe, or of a succession of universes, could not be such a cause, because it would itself, like Melchisedek, have no ancestry. Still, we may be told, the

Omnipotence

general relation of ground and consequent need not be confined to the form of it in which a particular finite cause has a particular finite effect. If so, the mistake in theological theory, if there be a mistake, might be simply that it spoke about a First Cause when it should have spoken about a First Ground. It would have taken too narrow a view of the logical relation in question. By taking an ampler survey of that very relation, we may be told, it might reasonably infer the existence of a first ground, a general determinable under which all specific finite events would be subordinated as determinates.

I hold (as I argued earlier) that this idea is untenable. We are talking about power, the power to produce, the power to persist, the power to induce changes. To speak of God as omnificent, is to speak of him as the source of all such power, as *being* all that power. Even if "power", in the sense asserted, did contain a common element with the general logical relation of determinable to determinate, that common element could not contain the specific character of power in its causal application. There may be what is sometimes called a "logic of events"[1]; and logic certainly applies to events. They and their relations contain nothing that affronts logic. But the general relation of ground and consequent, principle and instance, implicans and implicate, is timeless. As we have seen, temporal process may have certain properties to which time is irrelevant. There are timeless implications in process itself. The nature of a cause, however, is to proceed, and an account of causality that omits the *process* in it is inevitably defective in an ultimate metaphysical way. I cannot in fact believe that the process of becoming can be shown to be consistent with the immitigable inferential determinism of all future events that would be wrapped up in the idea of divine omnificence conceived in this way. By assuming that the future will resemble the past, and that every sort of transformation that will ever occur has already occurred, we might, indeed, as experimental determinists, proceed with our world-theories. That may be

[1] Personally, I don't believe that the phrase has any definite meaning.

Theism and Cosmology

the most promising policy to adopt, and I, for my part, would be quite prepared to adopt it as a policy. As a creed, however, it is blind. Even if the constant renewal of existence were of a slavishly repetitious kind, the renewal itself could never be a timeless logical implication. The logical inferences in question would be subject to an extra-logical (although not to an illogical) governing assumption or stipulation.

If, as pantheists say, everything is divine, all power is divine. If reality is a closely patterned unity, the power of it belongs to the patterned unity, and is divine if the unity be divine. These things may be true, and I am not attempting to dispute them. What I am arguing is that there is no constraining proof of the existence of an omnificent First Cause to be drawn from the attempt to amalgamate logical with historical process in the conception of causality.

Accordingly I submit that there is no clear, uninterrupted road from cause or power to omnificence in any of the three principal senses of cause. This conclusion, I daresay, may be generally received in an equanimous spirit. It is the fashion of the day to believe that school metaphysics usually attempted to demonstrate the indemonstrable. If the argument to an omnificent first cause is an instance, it would often be thought to be one of many. Again, the absence of a conclusive demonstration need not be a serious obstacle to belief. There is much that a reasonable man may believe although he cannot prove it to demonstration. If theists have to renounce the possibility of demonstration in such matters, they may still remain theists without the inconvenience of having to take leave of their wits. The failure of a demonstration is not a proof that there is not a power over the terrene scene with which pious men should endeavour to be in unison, or that this more than terrene power is not effectively supreme. It may be enough to reach the stratosphere without crying for the moon.

About this, I think, there is a good deal to be said. It should not be supposed that where many attempted demonstrations have failed, another may not succeed. On the other hand, the failure of serious and persistent attempts at demonstration is no sort of evidence that there *is* a valid demonstration hitherto undiscovered. If all the known roads are blocked, it is not impossible that access should finally be barred. Again, it is quite false to suppose that where there is so much smoke, there must at least be a tiny fire. It need not be so. If, for instance, a cause that is not also an effect is wholly unevidenced in terms of the uniformitarian theory of causality, it is absurd to say, that, in terms of that theory, there are probable but inconclusive reasons for believing in the existence of such a cause. Precisely the opposite would be true. The road would probably be closed.

On the other hand, it may be very unsatisfactory to distinguish three separate theories of causality, interrogate each of them, and leave the matter there. I shall conclude this lecture with some observations upon this complaint.

The three interpretations of "cause" or of "cause-factor" that I have discussed seem to me to be the chief and also to be distinct. If each of them were rejected, I do not see what would be left of causality; and their union, in some higher synthesis, in view of their many quarrels, could scarcely be effected easily.

It is not true, however, that the three theories quarrel about everything or that each fights in a separate ring. They may differ about the size of the ring, but they do have common ground. The type of fact that they discuss in its commoner instances may be illustrated by striking a match, by earthquakes, by the ruffled sea in a breeze. Where they differ is in the closer interpretation of such notorious facts; and in that they are similar to most speculative enquiries. To be sure, the debate may involve a certain revision of the territory that, *prima facie*, is discussed. That also is quite normal.

The common ground is or includes sequence, and the sequence, as I have said, has to be pegged down. In immanent causality it is pegged down to the same substance. In transeunt causality it is a transaction between specific things whose connection is not simply one of date. Some special sort of connexive continuity is always requisite.

This being granted, one question is whether the causal tie can be sensed or perceived, and very few philosophers have the effrontery to hold that we can see it, or smell it, or hear it. It is therefore different from change, movement or spaciousness, all of which are sensibly perceptible. The activists, however, maintain that the causal tie is directly experienced in volition. There, I think, they are wrong, at any rate if they hold that our acquaintance with our own volitions is a species of observation identical in kind with any other observation. What I think they do is to borrow illegitimately from the *logical* type of causal theory. There seems to be a definite logical continuity if we want to do X, set ourselves to do X, and then do it. If, however, we try to explain why we succeed, the argument is all the other way. We find by past experience that, setting ourselves to do certain things, we did succeed, and also that, setting ourselves to do other things, we did not succeed. We never know why; but is it plain that we have any other evidence?

With regard to the view that the causal tie is logical, certain general considerations must always be kept in mind. In the first place, cause and effect are correlative terms. Hence, given either, the other may be inferred; but, in that case, it has to be shown that any given event *is* a cause or *is* an effect. In the second place, everything is logical in the sense that it can be consistently described by means of propositions that contain general terms. That would be true of a chaos. The distinctive problem for this form of the causal theory is whether all existents have a causal structure that is of the order of a formal logical implication.

Here the principal question is whether time, sequence and process *can* be logicized to that extent and in that way. The order of earlier and later has a certain logical structure; but if the future is indeterminate before it occurs, it is not fixed before it occurs. If therefore it were implied by earlier process

Omnipotence

it could be implied only in a general indeterminate way and not in detail. It would therefore at the best be incompletely logicized. That would exclude omnipotence in its ordinary sense, although it need not exclude a theory of omnificence suitably circumspect in its limitation of all the power that there is.

The relations between the uniformitarian theory of causality, and the theory that causality is a specific logical nexus, may be complicated. The uniformitarian may hold, as Hume held, that unvarying sequence in the past and present does not entitle us to infer invariable sequence for all future time by any logical or logic-like principle. If so, he may also, like Hume, be an experimental determinist, that is to say, he may act and plan and calculate eclipses (say) on the assumption that the future will resemble the past with the nicest possible precision. Many uniformitarians, however, appear to hold that precisely the same events must always have precisely the same sequel, an echo of the logical view of causality meekly reverberating in a theory of mere uniformity of succession. This implies determinism, or M. Bergson's *Tout est donné*, at any rate if the universe be what is called a closed system. For, by hypothesis, the elements present at any time must be the very elements present at all times. It would not follow of course that we ever knew for certain what the cause-factors at any given time precisely were. Things that look the same, and have looked the same, for generations may not really *be* the same. There may always be secret or unsuspected differences. The theory is only that *if* they are the same they must issue in the same result.

As we saw in a previous lecture this form of the uniformitarian theory, or this alliance of the uniformitarian theory with an apparently straightforward logical inference, depends on the assumption that all process is at the most a transformation of the unchanging and that the transformation somehow does not count. I cannot believe that such a theory is ultimately tenable, and it may certainly be denied without absurdity.

VIII

Teleology

THE argument of the last five lectures has been concerned with empirically grounded theology in the widest sense of that term. The most general form of such an argument would be: "Something exists, therefore God exists", where the empiricism of the contention resides in the premiss: "Something exists." In comparison with such a very abstract empirical theology, the cosmological argument, in its usual form (that is to say the argument from the self-insufficiency of the *world*) is much less emaciated. It even relies upon a certain plumpness in the contours of what we call the "world", despite their ultimate self-insufficiency. The same may be said of the more special arguments we have recently considered, namely that passing existence implies eternal being, or as some would say implies immutable deity, that empirical existence is always originated and therefore implies an originating First Cause, that the scattered realities we encounter in experience, must be united in a ubiquitous divine spissitude; and the like. In these high regions, all the uplands are rather wide and rather bare. The empiricism of such empirical theology is therefore of a rarefied order; but it is much less abstract than the widest possible form of empirical theistic argument, namely that which begins with "Something exists".

In general, a philosophical argument is none the worse for being abstract, and if it depends upon an empirical premiss or premisses, there is no objection simply upon the ground that the "empirical" argument is less clotted with fact than is usual in empirical arguments. On the other hand, it must be confessed that most of us, even in our theology and in our philosophy, prefer, if we can, to be less resolutely and austerely abstract in our habitual thinking. To return to the metaphor of the uplands, we prefer the paths below the snow-line, and to be wayfarers in the middle which perhaps are the fertile regions.

That is one of the reasons why the Argument from Design has pride of place in so many theological discussions. There seems to be no humbug about it, very few pale abstractions. Paley and Bishop Butler tell us how we should argue if we found a watch on a deserted shore. We can see, quite easily, what they mean and can give our minds to such questions without taking fright at the outset, and concluding that the dangers of spiritual vertigo forbid us from attempting any metaphysical climbing at all. We are therefore inclined, rather too easily, to accept Kant's assurance that the argument from Design is "the oldest, the clearest and the most in conformity with the common reason of humanity",[1] and may even be prepared to argue against him regarding its shortcomings. Similarly in perusing Hume's *Dialogues* we are heartened to find that they are nearly all about the Argument from Design and that Philo, the sceptical interlocutor in them, seemed to find that argument so fit and so tough that after twelve rounds of in-fighting in the dialectical ring it remained as strong and as fresh as it was bewildered, showing to all the world that atheists are "only nominally so" and "can never possibly be in earnest".

True, we know well enough that these eighteenth-century ideas were not shared by the previous century at its metaphysical zenith. Descartes and Spinoza were theists, indeed Spinoza was God-intoxicated. Both of them regarded the design argument, in its usual form, as a vulgar superstition, drugged with the nonsense of final causes. In Mersenne's *L'Impiété des Déistes*, rather more surprisingly (since the author was a priest and an experimentalist) the good theologian sympathized with the deist in desiring an *a priori* instead of

[1] *Critique of Pure Reason.* "Of the Physico-Theological Argument."

a teleological proof of theism and so was willing to part company with St. Thomas and Aristotle. Our age, however, resembles the eighteenth century more than the seventeenth in many significant ways. It need not, indeed, be willing to accept what Dr. Tennant calls "the forcibleness of Nature's suggestion that she is the outcome of intelligent design"[1] and of the "wider teleology" that Dr. Tennant accepts. On the other hand, there would be a very general disposition to accept the view that if such a "wider teleology" were really forcible in its theistic persuasiveness, theism would be in accordance with the spirit of our favourite ways of thinking. To quote Tennant again, theism would be "reasonable belief" and would primarily involve little more than "the application of mother-wit to forthcoming facts".[2] It would be a revived form of the Argument from Design. The revival, moreover, is not confined to the present century. In the last century, for instance, Schopenhauer said that "every sound head must be brought by a contemplation of nature to a teleological position".

For these reasons I intend to give the Argument from Design as much space as I can afford—in all two lectures. In the first of them, I shall consider teleology in general. In the second I shall deal more specifically with design and with its relations to theism.

The present lecture, therefore, is preparatory to the next. In the next the examination of cosmological theism will be continued *sub specie finis*, but the present lecture is an interlude concerning the range and the meaning of teleological principle.

The most obvious instance of teleology or of final causality is the planning, scheming, striving and intending of conscious human beings. Thus ships and railways and aeroplanes are things that are devised. They come into being as means for human transport; and transport is something that human beings want for pleasure, or for business or for warfare.

[1] *Philosophical Theology*, II, 79.   [2] Ibid., II, 81.

Teleology

Proximately or remotely, such things are devised by human beings for human ends, and teleological human action of this type includes the vaguest as well as very definite schemes. One may want to write a novel and never get beyond the heading "Chapter I". One may also want one's lunch and want it now. Still less vaguely, one may want a Porterhouse steak at so-and-so's eating-house, and nothing else in the wide world. Teleological action, again, may be highly complicated as in the planning of a system of strategical railways, or it may be in appearance very simple, like a baby's impulse to suck.

I hold that in principle there is no mystery at all about final causes of this type. They are a species of efficient causes acting *a tergo*, and if any efficient causes are operative, causes of this order are also operative. What happens in their case is simply that the idea of something on which we are bent is a cause-factor in our actual behaviour. It remains before our minds until satisfaction is reached, or, alternatively, until boredom or dissatisfaction kills it. It leads to readjustments if we are balked but not dismayed, and may be very insistent even when the chances of success are obviously very slender. It is not invincible since it may be diverted by the intrusion of other interests or may elanguesce into nothingness; but it is often unconscionably stubborn.

This conception, I submit, is descriptive, accurate and straightforward. There is no occasion for playing tricks with it, as some philosophers do when they muddle the times in the affair and then regard it as a miracle. According to these gentlemen planned action must be action *a fronte*, that is to say, the future must act upon the present, and that (they say) is a mystery although it occurs so very often. Such a claim, however, seems to be wholly unnecessary. There is no need for anyone to believe that jam for tea at four o'clock makes our mouths water at three o'clock. What makes our mouths water at three o'clock is the thought of the jam we may have at four o'clock; and that particular thought occurs at three o'clock. The anticipations of future events occur before the future events, and may occupy us quite happily when they occur, even if the future, when *it* occurs, turns out to be clean contrary to the anticipations.

Difficulties, however, have been raised. There is a certain type of mind that is thrown into nervous convulsions when the most innocent statements are made about teleology. So I shall briefly consider two of the more serious of what seem to me to be hyper-sensitive apprehensions.

In the first place it may be argued that mental ideas cannot be the causes of physical actions any more than railway coaches can be coupled by the friendly relations between the engine-driver and the guard; and some philosophers would add (on grounds which must be quite different if they exist at all) that mental ideas cannot even be the causes of subsequent mental ideas in the same mind. If this were so our plans and schemes and other ideas could not act *a tergo* since they could not act at all. Therefore, if we believe we are actuated by such ideas, we must be mistaken. We should experience the drift of our appetites (as Spinoza[1] said) but we could not know the causes of our actions. Physicists and mathematicians, the anti-teleologists tell us, never make that mistake.

The people who argue in this vein profess to know quite a lot about causes. If they talk, like Clifford, about the train and the engine-driver and the guard, they may be invited to explain how the engine-driver reacts to the guard's whistle. Speaking as good physicists, they can scarcely maintain that the whistle starts the train. It is the engine-driver with his levers who does that, and there are surely very strong *prima facie* reasons for believing that the engine-driver's manipulation of his levers depends upon his ideas and intentions. Why should these *prima facie* reasons be jettisoned? If the contention were that pure disembodied thoughts on the engine-driver's part pulled the relevant levers, objections might proliferate very rapidly indeed; but the innocent description

[1] *Ethics*, Appendix I.

Teleology

of teleology that I have given above contains no such implications. I am not maintaining that the engine-drivers who manipulate their levers know precisely how they do it. They are not anatomists or physiologists. They need have no knowledge of the sort of jellification that (I am told) is the relevant physiological antecedent in the engine-driver's nerves. All that is maintained is that their intentions and decisions are among the efficient cause-factors in the case.

If that be denied the reason must be metaphysical or at least very highly speculative. It must be either that all mental processes are a description of what is inefficacious, or that such processes can have no *physical* efficacy. The former appears to be a mere dogma, the latter to depend upon a delusive revelation about the character of the causal nexus. This supposed revelation would be that mental events cannot ever influence physical events. Such an alleged intuition, however, is not a whit more plausible than any other among the professed intuitions of this order that we discussed and condemned in the last lecture. I am not asserting the utter disparity between mind and matter. It is Spinoza and his many followers, witting and unwitting, who did that. All I am saying is that human intentions and decisions appear to be cause-factors in human actions. They are among the best accredited instances of what we usually regard as a causal sequence.

The second supposed difficulty concerns value. Spinoza maintained, and many have agreed with him since, that a final cause is by definition a value-cause, or, as one might say in modern language, an axiological determinant. Such a cause, Spinoza went on to say, was wholly unknown to science; for science is always axiologically neutral. Consequently, if the hoary maxim *Quidquid petitur, petitur sub specie boni* were true it would necessarily be irrelevant to the order of efficient causes, and the inclusion of such an irrelevance would be disastrous to all clear thinking. The point, he continued, was peculiarly penetrating, for it concerned good and evil,

Theism and Cosmology

merit and sin, praise and blame, order and confusion, beauty and deformity. These were what is meant by value and disvalue, good and evil. None of them could be an efficient cause.

As it stands, this objection is irrelevant. The connection between teleology and value may indeed be very close, and certainly has a bearing upon the theme of the present lecture. We shall also have to discuss it, with a more extensive patience, in the second series of these lectures. If teleologists had to maintain that value, as such, determined existence, Spinoza would have made his point, and would have made a strong point. Teleology, in that case, would imply a type of explanation quite foreign to mathematics, physics or other such sciences. Even if Spinoza had been wrong in regarding such axiological explanations as indefensible, he would at least have shown that they should not be confused with the sort of explanation that he and his age regarded as rational or philosophical. No such objection attaches, however, to the view that conscious planning, or aiming, or scheming, may be an efficient cause. When considerations of worth or of value enter into such planning and devising it is the thought of worth and the belief in value that affect the issue. Why should these beliefs *not* be effective motives to human action as well as any others? Beauty may not be an efficient cause. It may not be capable of doing anything at all; but that is no proof that the thought of beauty, and the intention to enjoy or produce it may not passionately actuate a great many human beings. The moral rectitude of a deed may never be an efficient cause, but the belief in duty does seem to affect men's conduct.

Here, indeed, Spinoza's disclaimer seems to recoil on his own most admirable head. Logical validity is not an efficient cause. Logical correctness does not do anything. Nevertheless men have acquired a love for logic, a loyalty and devotion to it. If a man believes he can prove something, or disprove something, he commonly attempts the proof or the disproof. The desire to be logical, the belief that one can be faithful to

Teleology

logic, does move men to write and to debate. Such beliefs and desires actuate mathematicians and physicists rather more than most other people. They are an argumentative folk, teleological in their argumentative way.

I submit, then, that this second objection is irrelevant, indeed that it is irrelevant whatever view be taken of the status of value in the real world. Spinoza, in most although not in quite all of his pages, asserted that what we call value is merely subjective. It expressed (he said) the smoothness and the sweetness of mental movements. It was not a natural property of natural things. We may reply that if this were the truth, such subjective feelings might nevertheless be cause-factors in the actions of men and women, that is to say, in the actions of the persons who have these private feelings. If there were no beauty except in the mind of the beholder, if beauty were simply a name for the fact that certain things evoke a peculiar delight in men of culture, this delight and the lingering rapture of such men for the cause of their subtle enjoyment may well affect their actions, leading them to paint, or to haunt the galleries, or to write articles for the press. The subjectivity of subjective causes is in no way a barrier to their efficacy.

Accordingly I shall assume, for the rest of our discussion, that consciously intended actions do occur, that there is no sufficient reason for denying that in such cases the intention is a cause-factor contributing to the result, and so that, in that sense, the existence of teleological behaviour in the world need not be in dispute. Such behaviour is idead and purposive. It therefore pertains to beings who *have* idead purposes, in such part of their behaviour as is idead and purposive. What has now to be asked is whether idead purposive action of this type exhausts all the teleology there is *in* the world, and also the teleology *of* the world, if such teleology there be.

The words I have used up to the present, such as "planning", "scheming", "devising" and "intending" implied the presence of ideas of a pretty high grade. It seems to be perfectly clear,

Theism and Cosmology

however, that we should be very loth to restrict teleological behaviour to idead action of that high grade. Even those of us who have attained years of discretion do not always act with clear intentions. Our appetites are frequently purblind, tentative, troubled and hazy. We suppose that babies, and cats, and owls and puppies and sparrows behave teleologically although they cannot entertain clear, or general, or explicit ideas. They seek or strive although they do not plan or devise. Cows, as the Indian philosopher said, do not intend to produce milk for their calves. At any rate we think they don't. In other words, we assume that idead purposive human action is a subtle and elaborate refinement of a much more extensive type of behaviour, and we think of teleology as descriptive of the general type, not simply as the name for its idead refinement.

Plainly, there is force in these contentions. There are low-grade as well as high-grade ideas, and the lower limits of idead teleological action may be very difficult to define. That, however, is not a proof that there are no such limits.

One suggestion is that all *interested* action is teleological (to borrow a useful term from Mr. Perry's *General Theory of Value*). "Interested" action, in this sense of the word, is any action that rouses or is warmed by any interest, any action that is pursued with zest or is anguished if it fails. It is action that is *felt* or at least is capable of touching the quick of our feelings.

In that case, a very definite problem emerges. In idead teleological action there is a definite cause-factor that can be shown to exist and may reasonably be presumed to be operative. This is an expectation of the sort that, if we reflected upon it, we should deem to contain the promise of some positive value. The question now is *what* antecedent can be shown or presumed to exist in interested conations that are not explicitly idead in the same simple sense.

Consider, for instance, a very familiar problem. When we have past experience to draw upon, it is plausible to say that,

Teleology

having found we can do certain things if we set ourselves to do them, and having found the result satisfying, we attempt to do them again expecting a like satisfactory issue. This type of explanation may be extended to hazardous enterprises that have only a generic resemblance to anything we have attempted before, hope rather than knowledge being our star. The assumption, however, is that such an action is not wholly blind either as regards the end we hope for or as regards the means that we expect to be an efficient preliminary towards that end. That being assumed, what are we to say about the *first* teleological action of the kind, say about the first performance by an individual creature of the sort of action that is instinctive in its species?

One of the better known definitions of instinct, it may be recalled, is that instinct is "the faculty of acting in such a way as to produce certain ends without foresight of the ends and without previous education in the performance". These were the words of William James[1] in a discussion of the subject of instinct that did much to stimulate the immense vogue of instinct-theories in Edwardian British psychology. (The vogue is fading now.) James's definition is positive as regards the teleological character of instinct even on its first performance, but sharply denies its identity with idead final causality in the way in which we commonly think of that process. Our problem is how such teleology should be conceived where no ideas regarding the end or the means can be assumed to be due to individual experience.

Certain psychologists hold with Stout that the distinguishing feature of instinctive action is the possession of a rather elaborate inherited mechanism of response adequate for certain situations such as being afraid, being thirsty, or being amorous. That would explain why the appropriate means are ready for use, but it would not explain why they should be used, and timed, for the appropriate end. Others invoke a racial or inherited memory, thus pushing the problem further

[1] *Principles of Psychology*, II, 383.

back but not solving it, and, in addition, treating the term "memory" in a very high-handed fashion. A third group seems to assume a divination of the end on the first performance, the divination being described by apparently humbler terms such as "intuition" or "sympathy". In the opinion of this group, the moorhen that makes its first dive when it is startled by a puppy divines that it will be safer in the water than on the bank. This third theory would amalgamate such teleology with idead teleology. It would simply say that the moorhen was a prophet. But it is very hard to accept such a doctrine.

All these theories, in short, appear to be either insufficient or unconvincing, and I do not know of any others that are better. Each of them however makes a comparatively candid attempt to face a genuine difficulty, and in that important respect each of them is superior to a shoddy device that is sometimes used, the subterfuge, namely of saying that men and other animals, in the first performance of an instinctive action, behave *as if* they were pursuing an end, although strictly they are *not* pursuing an end. That is a mere evasion of the issue, even granting that Kant himself talked about purposiveness without purpose, *Zweckmässigkeit ohne Zweck*.[1] The question is whether the category of end enters or doesn't enter, not whether there may not be a certain resemblance between behaviour, on the one hand, in which an end is foreseen and pursued, and, on the other hand, behaviour in which no such end should be supposed to be *foreseen*. Certainly there is a resemblance, but that is precisely where the anti-teleologists triumph. There are plenty of instances, they say, which *look* as if they were planned, but turn out, when we examine them closely, not to be planned at all. Therefore I submit that the great "As If" is only a subterfuge, unless it is regarded, not as a solution, but as a statement of the problem.

As it seems to me, the trouble about all "as if" philosophies is that they stop too soon because they are far too easily tired. What they assert in their languid way is usually defensible

[1] *Critique of Judgement*, § 11.

Teleology

There are analogies to be explored, but instead of exploring such analogies the "as if" philosophers simply display them rapidly, and, like a cautious auctioneer, decline to guarantee the genuineness of the goods they have to sell. That may be better than mistaking analogy for identity, but it is not nearly enough, and the obvious remedy is to examine the analogies closely, pointing out with precision what the resemblances are, and where they end, where the differences begin, and what the chief of them are. That is what we should do if we hold, as we may reasonably hold, that interested behaviour may *resemble* idead purposive behaviour without itself *being* idead. Having indicated the remedy, therefore, I shall now attempt to prescribe it.

What happens in idead teleology? The answer is that an expectant or prospicient idea moves and guides the ensuing action, being a part-cause of that action. The further explanation that is usually given is that such ideas move us because they are ideas of value or of benefit. This additional explanation, we may allow, is probably much more intricate than it seems. We have to ask "Whose benefit?" and "What sort of benefit?" We have also to ask whether it is seeming benefit or genuine benefit that moves; whether the idea of benefit is not rather a reflection of incipient appetite than, so to say, an original appetizer; whether what is sometimes called the "true" end of an appetitive process may not be masked by ends that are merely specious. All these problems, and many others, cluster round the assertion. They are heavy with the perplexities of value-theory. Despite these perplexities, however, we may confidently assert that idead teleological action, at any rate in the normal case, is action under the guise of apparent benefit, action *sub specie boni*.

This being allowed, we may reasonably concentrate our attention upon prospicience and upon value when we attempt to compare idead teleology with teleology that is interested but, in strictness, unidead. As regards prospicience we may further reasonably suppose that the only sort of foresight that is possible is conjectural anticipation by means of ideas. There is no literal foresight of the future. It follows that in unidead teleology there is no prospicience. For all conjecture is idead. Here the word "idea" includes vague ideas, and indeed must include anything that may fairly be called conscious expectation on the part of any animal as well as the general and explicit ideas that are characteristic of human scheming. I refuse to believe, however, that the startled moorhen making its first dive has any ideas at all, however vague, regarding the benefits of water as a medium for escape.

Accordingly the conclusion is that in idead teleology there is prospicience, and that in unidead teleology there is not prospicience. In the former, a conjecture or expectation regarding the future operates as a part-cause in the present. In the latter, properly speaking, there is no conjecture and no expectation. Nevertheless, although there cannot be *prospicience* in unidead teleology, there may very well be preparatory adjustment, *pre-adjustment*, and this pre-adjustment may be highly specific as in the instance of the moorhen's first dive.

That I suggest is the proper conclusion of the matter so far as the mark of prospicience is concerned. Unidead teleology is often preparatory but not prospicient. It is often pre-adjustment that is not, in any intelligible sense, "prospective". It resembles idead teleology because pre-adjustment resembles the special sort of pre-adjustment that is really prospective. That is where the great "As If" comes in. But there need not be any prospective ideas, since the preadjustment may be quite unidead.

This distinction, it is true, may seem to be niggling or obscure, so long as we confine our attention, as at present, to interested action; but, as we shall see, it becomes obvious and inescapable when we examine certain biological pre-adjustments in which there should be no question of feeling or of interest, to say nothing of ideas. There is vegetable pre-adjustment when the buds appear on the trees in spring, and if, to use one

Teleology

of Mr. Wells's phrases, we are to account for the arrival of the fit as well as for the survival of the fitter, it may well be necessary to postulate some vital pre-adjustive *élan*.

But let us return to our general topic. Prospicience and value are the two marks on which we have agreed to concentrate our attention. What, then, of the second mark, the mark of value?

I have spoken about idead action *sub specie boni*, but it is plain that we cannot suppose that a precise or elaborate "idea of value" is present in all such action. The man who does his duty for duty's sake may have such ideas. So may the logician who devotes himself to the straightest of straight thinking. So may art critics, although very seldom artists. If, however, we say, in the common way, that Mr. Everyman invariably acts *sub specie boni* in so far as his behaviour is idead at all, we cannot suppose that he is a philosopher and a theorist in all his actions. What must be meant, therefore, is (at the most) that if Mr. Everyman reflected upon his idead actions, or, more probably, if he got an expert philosopher to reflect for him, it could be shown that he did seek what appeared to him to be a good. The philosophical experts whose services were thus employed would differ profoundly and with some acrimony regarding the fuller meaning of this statement, and some of them would renounce it altogether, but a large proportion of them would advise Mr. Everyman that, in some significant sense, the general statement was probably true.

In other words, the ideas that, according to the schools, "seem good" to Mr. Everyman and therefore move him to act, are not usually *about* goodness and seldom include analytical attention to the meaning of "good". That is very clear in most of the traditional, and in many of the modern, accounts of action *sub specie boni*. Freudians say it is a wish-fulfilment that may be predominantly or wholly "unconscious". Hedonists say that it is action in which pleasure spurs us on and pain makes us shrink and flee, but (if they are at all astute) they do not assert that we are always thinking about pleasure, or calculating it with what Benjamin Franklin called a sort of "prudential algebra". Opponents of hedonism, such as the Stoics and Bishop Butler, maintain that we are moved by certain instincts or other primary dispositions whose function, *de facto*, whether we know it or not, is self-preserving, or race-preserving or sociable in some other way, pleasure or pain being only a sort of register of the success or failure of these primary appetitions. According to all these theories of action *sub specie boni* the general, abstract, analytical idea of "goodness" is seldom present and is hardly ever operative. That is essential to the theories, unless they are prepared, quite obviously, to desert the facts. For human beings do not generally philosophize about goodness when they act.

For the most part, when it is said that human beings act *sub specie boni*, the meaning is only that human beings tend to do what attracts them and tend to avoid that towards which they feel distaste or repulsion. What "seems good" to a man is what attracts him and the "good" or attraction is regarded as a pleasant feeling that attends or suffuses certain of his anticipations and is ostentatiously absent from others. There may be a deep natural reason for the stronger and commoner among such attractions. Jack's interest in Jill, for instance, may be Nature's way of facilitating a further supply of little Jacks and Jills; but the superficial motive, even if it be little more than superficial, is the feeling of attraction, or, as Mr. Perry would say, of "interest".

"Interested" action may therefore appear to provide a natural bridge between idead and unidead teleology. It occurs in idead action; for we do have expectant ideas when, in the one case, we are attracted, and in the other case we are repelled. The feeling of attraction, on the other hand, is not in general an "idea", but is only a timbre of feeling that tones or tinges an expectant idea, or, more generally, a preparatory state. It may be made the object of reflective contemplation when we philosophize about it. In that case (many philosophers would say) it provides the basis for the philosophical idea of "good-

Teleology

ness" if, indeed, it does not provide the whole of that idea; but in its primary occurrence it is not an idea. Again the further inference is sometimes drawn that since most natural appetites and other attractions are biologically useful (hunger, for the most part, bidding us eat as we should, and thirst, again for the most part, bidding us replenish our de-hydrated bodies in a useful way) these attractions and appetites are usually "good" in another sense. They lead in the main to health and strength and a longer life; that is to say, they are beneficial in what is usually supposed to be an objective way.

I cannot believe, however, that there is any natural bridge of the sort alleged. This entire theory of action *sub specie boni* refers to the attractiveness of certain prospicient ideas, and asserts that such attractiveness is either what we mean by "seeming good to a man" or is a natural vaticination of human good. The fact that other, more analytical, more reflective and sophisticated ideas of value may or may not be present is, so far as I can see, a simple irrelevance in this connection. In other words, the *appearance of good*, if there must always be such an appearance in prospicient teleological action, does not even begin to unite unidead teleology with idead.

In idead teleology there is a definite efficient part-cause, viz. the prospicient idea or expectation. This part-cause may always be felt as attractive (when the action is *towards* not *fromwards*), and the fact may be of great importance for value-theory. But no more than this can be inferred. In unidead teleology (if there is such teleology) no such cause-factor is to be found. As I have said, there may be a state of pre-adjustment. That, however, is not a distinctive part-cause. Instead it simply gives a proleptic statement, like the statement that a bomb is about to explode.

Idead teleology, therefore, differs characteristically and most significantly from unidead, and it is merely evasive to cloak the matter gracefully with prattle about "As If". But unidead teleology undoubtedly occurs. The marks we attribute to it are marks that characterize certain patterns of natural

Theism and Cosmology

behaviour. They are, for example, characteristic of all organisms. We need not suppose that either oaks or lime-trees expect the winter; but they do prepare for it. They have also a nisus towards self-preservation and towards the reproduction of their kind. Again they advance towards maturity. Speaking generally, the selection of what in some sense is beneficial, and the *utilization* of the environment, are patterns of process in all living things. Organisms adapt themselves, make use of other things, extract benefit from their surroundings. In other words, unidead teleology is a fact in nature, and occurs where there can be no question of ideas, wishes, decisions and design.

In short, "interest" is *not* essential to unidead teleology, and the term "interest" in the special sense in which Mr. Perry and others employ it is at least quasi-psychical. As I have said, "it must at least be capable of touching the quick of our feelings". It is applicable, therefore, to all living beings who can suffer; and there are very good reasons for believing that the pains and the pleasures of the higher animals correspond fairly closely although never quite precisely with important evils or benefits to these animals. Even in mankind, however, there are many vital matters that do not touch the quick of our feelings; and much that does touch the quick of our feelings is *not* vital. The early stages of a cancer need not be felt; the pain of rheumatism need not be a serious menace to our prospects of survival. Much that is essential to health, for example the presence of vitamins, has been unsuspected in former ages, and has no direct connection with pleasure-pain at the present time. Accordingly if we rid ourselves of the preconceptions of ideed teleology when we are thinking about unideed teleology in nature, there would seem to be no occasion for supposing that teleology implies "interest" of a psychological or near-psychological kind. To suppose so is vastly to exaggerate the biological importance of the central nervous system. Adaptation, co-adaptation, self-maintenance, reproduction and the like are the "ends", often quite unideed

Teleology

and quite unfelt, of the sort of biological pattern that we call teleological. They describe certain patterns of natural fact in terms that imply neither the ideas nor the feelings of any mental being. They are very much wider than "interest" in the sense we have chosen to give to that term.

The reason, I think, why so many philosophers are so reluctant to admit the reality of non-mental teleology is that teleology seems to them to be inseparable from some kind of value, and that they hold that nothing except a state of consciousness can have value, except in the sense of being a means to or a condition of such a value. Sometimes these philosophers regard this contention as axiomatic, guaranteed by a pellucid intuition. Sometimes they say, more warily, that it is the inescapable result of an empirical survey of values. They *see* that states of consciousness are good or evil and they *say* that they cannot conceive of anything else that could have value or disvalue. Hence they infer that if there be no "interest" at all there cannot be teleological action. For them the point at which *psychical* "interest" ceases is the point at which "value" ceases.

Here the dispute is, in part, about words. If we mean by a "value" that which, as a man would say, makes *his* life worth living, then the vegetable teleological patterns of which I have been speaking, such as self-maintenance, development towards maturity, or the perpetuation of the species might be denied value of every kind. I say "might" not "must", because many men have said that existence, as such, is better than non-existence, that agony and madness are preferable to death, and, I suppose, that if reincarnation were the truth, and the fate of a man were to become a cabbage, it would be better to be a cabbage than to be nothing at all. If so there might be some comfort in the reflection that even a man's body cannot be wholly destroyed.

I do not know what should be said about such views. If a man says, like Milton's Satan, that hell itself would be better than extinction he may be presumed to believe that in hell he would retain the consciousness of his own identity. He would still have a mind, and so would be privileged despite his grievous condition. In short, our imagination is apt to flutter unsteadily when it takes wing in these insubstantial regions. What seems to me to be plain is that self-maintenance, development to maturity, the avoidance of mutilation (and the like) are among the *ends* of all biological existence, whether in man or in any other living being, that life at any given time is a preparation for continuance, and that such patterns are teleological. It does not seem to me that life, in sensitive and in intelligent creatures, can be said to be a mere means to agreeable sensations, or to fine feeling, or to high thinking. It is the living being that is happy or intelligent, and the substance of a thing cannot be a mere means to any of its states. Consequently it seems to me to be false to say that, even in mankind, biological ends are only means to meta-biological values. Damage and pain are not identical. Pleasure and health are not identical. If it be said that the mere continuance of existence need not be valuable, the question is what is meant by such "mere" continuance. The meaning should be that a man continues to be *himself*, that a horse continues to be one and the same particular Dobbin; and so forth. I cannot see why maintenance-values, in that sense, should be denied to be "values" in all relevant senses of the term, and I propose to regard them as genuine values.

Therefore I think we should deny that value, in all its senses, is inapplicable to what is non-mental. The alleged intuition, to the contrary, is a covert assertion (I believe) that value is always a kind of satisfaction, and that every satisfaction must be felt. Opposing this view, I am disposed to hold that satisfaction, where it is felt, may often be relatively superficial, and that among living creatures the furtherance or obstruction of teleological patterns is usually a much more serious matter. If, then, the alleged intuition is abandoned, the arguments that are drawn from a humbler empirical survey cease to be formidable. Without the alleged intuition they are quite forlorn. Let us grant that human beings, reflecting upon the values that *they* may achieve, find that, in their case, high thinking, subtle artistry, moral integrity, the gusto of rivalry and the zest of sensuous joys are so incomparably superior to all other human values that nothing else seems to be of any account in their lives. Does it follow that there are no values in sub-human beings, and that other living things, successfully following the bent of their natures, selecting what is favourable to the pattern of their existence from their environment, using that environment to further their own entelechies, are not, clearly and quite objectively, in the same predicament as mental beings who have mental values? That is not a matter of words, but a matter of fact. If the term "value" be so defined as to exclude maintenance-values, some wider term would have to be found to include them both.

Mr. Perry defines "interest" as a "permanent embodiment of motor-affective bias".[1] It is restricted, therefore, to feeling and to movement, i.e. to feelings and to the sort of movements that we call reflexes, conditioned reflexes and voluntary movements. In this sense, reflexes and conditioned reflexes would not seem to tally very closely with his tale, but if they did, motor-affective "interest" would be a mutilated description of the facts of biological teleology. It would omit the patterns of growth, development and organic regulation that were not affairs of muscular movement, except in so far as these were attended indirectly with conscious feelings. It might admit growing-pains, but not growth itself. There are patterns of absorption and secretion, teleological in the same general sense as gross muscular movements. The body of a living man is a natural thermostat and a natural calorimeter as well as a muscular agent. These functions, it is true, have an effect upon our feelings, on our moods of depression or of ebullient spirits; but their vital worth is immensely deeper than any such feelings, and is not an affair of motor or muscular response. The teleological pattern of vital existence, in short, is mis-

[1] *General Theory of Value*, p. 27 n.

construed if exclusive emphasis is laid upon its muscular behaviour-pattern, and the point has importance, not simply in itself, but also because it exaggerates the resemblance between idead and unidead teleological process. In the idead way, we pursue our purposes by voluntary movements, snatching our prey and masticating it. What happens afterwards is not an affair of volition, but it is an affair of growth and of vitality.

Hence, I submit, we should infer that teleological process in living beings is much more extensive than interested behaviour of a "motor-affective" type, and, *a fortiori*, much more extensive than idead choice. This seems to me to be true with respect to both of the primary marks of teleology that I have discussed, viz. specific futurity-adjustment or "preparation", and maintenance-value, the *conatus perseverare in esse suo*. Such futurity-adjustment is a purely descriptive term and does not connote even the faintest traces of prospicience. Again, I would not shrink from applying the term "value", in the sense of maintenance-value, to a descriptive biology of sea-anemones and of honeysuckle without any implication of ideas, purposes, "interests" or pleasures.

Living things are the most obvious and the most striking examples of unidead teleological units in nature. I have therefore used a good many biological strokes in the sketch I have attempted to make of their teleological pattern. It seems clear, however, that the sketch, if there be truth in it, should not be confined to biology. I do not mean simply that the boundaries between living matter, and matter where life is absent, may be rather hard to draw. "If into this beaker, containing a super-saturated solution of sodium thio-sulphite" says Mr. G. N. Lewis,[1] "I drop a minute crystal, it *grows*. I take a bit of this new growth and inoculate a second beaker: and so I may go on." Such statements may show that one of the primary characteristics of biological entities, namely their capacity for growth, has a near-exemplification if not a com-

[1] *The Anatomy of Science*, p. 179.

Teleology

plete exemplification in chemical entities that are not alive. That is an important matter, but I want to advance to an assertion of still greater importance. As it seems to me, a closer analysis of teleological pattern yields the conclusion that teleology in nature, of a wholly unidead kind, is not confined to biology or to the near-biological. In biology, reproduction, patterned growth to maturity, and the assimilation of foods that favour such growth are the distinctive features of the pattern. Futurity-adjustment and the selection, or utilization, of what is favourable to maintenance-value are the general characteristics that are shown, in a special form, in the distinctive patterns of biological reproduction, assimilation and growth. These general characteristics need not and should not be confined to the form that they evince in biology.

What we call adaptation or co-adaptation is the same thing as specific futurity-adjustment though rather wider; and co-adaptation occurs extensively in the world. Selective utilization, again, occurs very frequently indeed, and need not be confined to the instances in which a man makes use of a Daimler car, or a fern makes use of the water near its roots.

I shall now try to develop these ideas.

Consider, firstly, adaptation. It must be conceded that when we think of adaptation, we usually make a latent reference to conscious planning and design. Things, quite obviously, are adapted when they are *made* to fit, and they are made to fit, we say, when somebody plans the fitting. Such statements, however, obviously ignore the primary fact that when things are successfully made to fit, they really do fit. The fact that they do fit is what I am calling a teleological pattern. It is because there are specific and close adaptations in nature that we are able, in so many ways, to use nature for our human conscious purposes by taking thought, and by altering the face of nature in accordance with our thoughts, and schemes and desires.

People talk about mechanism *versus* teleology. In doing so they seem to forget what a machine really is. It is an artificial

Theism and Cosmology

teleological entity, a physical thing so adjusted that it achieves certain results, printing the news upon the tape, adding a bank balance from day to day, registering cash, or whatever the job may be. The electric hare is as well adapted to greyhound racing as a living hare, perhaps better adapted. Machines do not merely present the illusory appearance of being adapted to the achievement of certain results—to fly, to count, to sail the seas or the like. They *are* so adapted, and although I have, for the most part, selected illustrations in which a designing mind is intermittently affecting the machine, my remarks would apply equally to machines that did their work while we were all asleep. Such machines have no teleological ideas, for they have no ideas at all; but the pattern of their action is teleological.

If it be replied that machines do not spring up of themselves, but have always a mind that is responsible for their origin, the answer is that this may be true about railway engines, electric hares and other such machines, but that the fact may well be irrelevant. Such machines are made out of unpromising materials where nature has been niggardly to man. The special combination of steel, copper and the rest that we call a railway engine, does not turn up without human scheming. If a stream-lined Royal Scot were excavated in a new coalmine in Manchukuo, the thing could be quite as astonishing as the legendary monkey at the typewriter that was discovered to be tapping out the story of *Trent's Last Case*. *Per contra*, it would not be at all surprising if the fossil remains of a tree were dug up in the mine. Why? Obviously because the sort of teleological being that we call a tree does turn up in nature without human contrivance, and used to exist *in situ* at about the time the mine was naturally formed. Organisms are just the sort of teleological beings that we do expect to find in such places. Stream-lined railway engines are not. But the constituent parts both of the tree and of the engine are mutually adapted in a teleological way.

To be sure, it may be objected that the machines that are

Teleology

designed by human beings are designed for human purposes, and that they are teleological simply in the sense of being serviceable to human beings. That would be an extrinsic teleology of these machines, and if the teleology of anything other than man himself is regarded in this extrinsic fashion, there is no problem, and no reason, except negligence, for supposing that there is a problem. In this extrinsic sense, anything other than a man is teleologically related to a man if it is of service to the man, if he can use it, and derive some benefit from it. Design need not enter. If the environment is propitious it is of service whether its services are or are not cunningly obtained. Berries and nuts have this sort of teleological property as well as baker's bread and the command of the seas jealously guarded by a nation that cannot grow enough wheat for itself. What I am discussing now is not the extrinsic teleology of being serviceable to mankind or to some particular man or nation or sect, but the general character of teleological pattern, extrinsic or intrinsic. From the extrinsic standpoint fish are serviceable to men, and worms to fish, and men to worms when the men are buried. The relation seems to be fundamentally the same in all three cases, for I have already shown that it need not be confined to the services that are done to human beings. Intrinsic teleology, on the other hand, refers to the reciprocity of the constituent members of an entity which is prepared or ready for a futurity of self-maintenance, short or long. In this case also I have argued that such patterns are frequently to be found in nature, and may also be artificially produced. They are sub-mental in many animals and in all plants, but they are not confined to living beings. What is more, it would not seem to be impossible that Nature itself at any given time, if it be a whole, is a teleological whole in this sense, and so that there is a teleology *of* Nature, as well as teleology *in* many natural things.

That is the story of teleological adaptation. More particularly, the mutual services of the co-operative parts in any intrinsically teleological pattern, as in the classical fable of

Theism and Cosmology

Menenius Agrippa regarding the belly and the other members of the human body, illustrate co-adaptation, and so does the extrinsic teleological relationship of using and of being used. On the latter point, some further discussion would seem to be appropriate.

"The eye", said Paley, "is made for light, and light for the eye. The eye would be of no use without light, and light perhaps of little use without eyes."[1] The good archdeacon's hesitation, in the last part of the sentence, is interesting and rather significant. We may pursue the point with special reference to organisms and their environment, although, as I have said, the general problem is very much wider.

Many of us would hesitate if we were asked to assent to the statement that an organism can be said to be adapted to its environment in the same sense as its environment can be said to be adapted to the organism. In a manner, of course, the thing is past disputing. If an organism is suited to its environment, the environment is suited to the organism. But if the organism uses its environment, all that follows is that the environment is used by the organism, not that it is also a user or has any reciprocal function of that order. Indeed, we should usually recoil from this inference, and accumulate contrary evidence. The parasite uses its host, but the host does not use the parasite. There is one-way comestible traffic when the spider devours the fly. The spider benefits and the fly loses its all. The moth, we should say, is drawn and tempted towards the candle, but the candle is not drawn and tempted towards the moth. There is non-indifference and a natural egence (as Grote called it) on the moth's part, disastrous to the moth. On the candle's part there is complete indifference.

Co-adaptation, therefore, may seem to be an odd word for describing this matter. There is of course, no contradiction when it happens to be true that A makes use of B, and that B, also, makes use of A. When this happens, however, the uses would seem to be different uses. The postman makes use of

[1] *Evidences,* chap. xxiii.

Teleology

the nation as the source of his pay, and the nation makes use of the postman for delivering its letters. Even if the postman is also a taxpayer when he buys his tobacco, there is a clear distinction between his tax-paying capacity, and his reception of the social services dependent on taxation. In short, as we have seen, if A uses B, B is used by A and that is all the reciprocity that the case need contain. It is the reciprocity of translation from the active into the passive voice, a thoroughly attenuated kind of reciprocity.

That is quite true, and it is also quite innocuous. The relation of using and of being used does not apply to intrinsic teleology, for it is only by metaphor that anything can be said to make use of itself. When, in extrinsic teleology, it makes use of something else, the relation need not be symmetrical in respect of self-maintenance, benefit or advantage in the widest possible sense that can be given to these terms. In general we think of the user as in some way "higher" or "better" than the thing that is used, but that would not be true in the case of most parasites. I do not think that any further discussion is required.

I shall now attempt to give a free rendering of the main contentions I have tried to bring forward in this lecture.

It is often supposed that teleology is a mad intruder into anything that purports to be a scientific description of events in Nature. Kant called the principle a "stranger in natural science"[1] despite his eagerness to find some sort of philosophical justification for it. For many centuries there has been a persistent agitation to have the "stranger" classified as an undesirable alien.

The principal reason for the agitation is the general presumption of modern philosophy that the only possible cause is an efficient cause, or, as one might vulgarly say, that a kick in the pants is the only intelligible type of causal relation. I am not attempting to dispute this idea of causality. If anyone, at the present day, tries to revive the αἰτίαι of Aristotelian philosophy, say the "Formal Cause" or the "Final Cause"

[1] *Critique of Judgment*, § 72.

Theism and Cosmology

of the Stagirite,[1] he is talking about principles or aetiologies that are not "causes" in the sense in which we are familiar with that word. What I am saying is that if by a "cause" one means an efficient cause, in the vulgar as well as in the scientific and in the philosophical way that is customary to-day, then the only final cause that is an efficient cause is an idead cause, that is to say a witting decision, or some other such idead cause-factor. I am not attempting to discuss how big a part such causes play in the world's affairs. They may often mask unwitting tendencies and what Freudians would call "unconscious wishes". I am saying only that these idead efficient causes do seem to be operative, if anything is operative; but I should also say that our docks and railways and mines and agriculture would not be what they are without a great deal of conscious devising and of thoroughly idead scheming.

In these cases we can point to a definite conscious antecedent of a result that is frequently successful, and this antecedent is an efficient cause. It is also quite likely that what I have called "idead" action may be effective, rather widely, among beings whose "ideas" are not of a very high grade. Wherever there is prospicience in alliance with appetite there would seem to be an efficient cause of the type I have indicated. In much biological process, however, whether in men, or in the other high animals, or in plants, there would seem to be *no* prospicience, and *no* ideas, although there *is* teleology.

Such teleology is analogous to idead teleology in sundry ways, but it is not efficient causality, because there is no assignable antecedent that, like the idea of future benefit, influences present action. In its case, what we have to note is that there is a pattern of growth or of movement that may be described as a preparation for specific futurity, and as the tendency to achieve some benefit, if only a benefit of the sort that may be called maintenance- and development-value. Neither value nor the future is an efficient cause. The idea of

[1] e.g. J. L. Stocks in *Time, Cause and Eternity*.

Teleology

value and the idea of futurity are efficient cause-factors, but not the value itself or the futurity itself. We have therefore to abandon the idea of a teleological *cause* in all such instances, but we do not have to abandon, and we should not abandon, our beliefs in a teleological *pattern* of growth or of movement. Such teleological patterns do occur in nature. They should not be "strangers" to science or to philosophy. They are descriptive conceptions, significantly applicable to a host of facts that cannot reasonably be denied. As such they are not at all subjective. They are not just the shadows of anyone's mere opinions, caprices or point of view. Whether they are or are not useful in any science or in any philosophy may be largely a question of policy. The policy of teleological science in biology may or may not pay, or if it pays in certain scientific eras it may not pay in others. I am not, at the moment, discussing the fruitfulness of the notion. I am only defending the truth of it.

It seems to me to be a profound mistake, both in principle and in policy, to attempt to amalgamate idead with unidead teleology, because of certain resemblances that may be discovered in the pair. Idead teleology contains an assignable efficient cause. Unidead teleology does not. Thus the attempt to combine the two in terms of a theory of motor-affective "interest" seems to me to be conceived on the wrong lines. It is definitely misleading because it would renounce the name of teleology except where there is something psychical or quasi-psychical, intelligibly to be described as an "interest". The renunciation is gratuitous. Growth and development, the most fundamental notions in biology, need not be idead and need not be sensuous; but they *are* teleological.

If the concept of teleology stopped with living things it would still be a notable principle. I have tried to show, however, that it need not be restricted to what is alive. Adaptation, co-adaptation, and the selection of what is favourable to self-maintenance imply a teleological pattern and are principles that extend much more widely than biology. Biological

Theism and Cosmology

patterns would seem, indeed, to be a special case of this more general relationship. When we think of organisms we think of something that is *organized*. Their biological organization does not imply an organizing idea or purpose either on their part or on the part of anything else. It is a name for the fact that they *are* organized. And organisms are not the only natural entities that are organized. A pattern of organization is a teleological pattern, and is exhibited wherever there is adaptation, co-adaptation and using.

Dr. Whitehead[1] has recently called attention to a passage in Bacon's *Silva Silvarum* which begins as follows: "It is certain that all bodies whatsoever, though they have no sense, yet they have perception; for where one body is applied to another, there is a kind of election to embrace that which is agreeable and to exclude or expel that which is ingrate; and whether the body be alterant or altered, evermore a perception precedeth operation; for else all bodies would be alike one to another". I would not myself speak of "perception" in this connection, or speak of what is "agreeable" or "ingrate", although I would not deny that the pattern of what is mind-like, as well as the pattern of what is life-like, may be much more widely diffused in nature than we are usually inclined to suppose. What Bacon called "the election to embrace" and what, in another book, I have called "natural election",[2] is not (I have argued to-day), either idead or "interested" in the sense in which we commonly and reasonably use these terms. Nevertheless natural election, or the principle of selective *use*, does appear to me to describe an immense range of natural fact, and to be the fundamental conception in all unidead teleology. That is the type of teleology to which I have paid most attention in this lecture.

[1] *Science and the Modern World*, p. 52. [2] *The Idea of Value*, chap. iii.

IX

The Argument from Design

THE principal theme of the last lecture was particulate teleology, the adaptation of this or of that within the world to this or the other particular partner. The traditional argument from design, in Boyle Lectures and in Bridgewater Treatises, has usually been based upon these particulate adaptations. Indeed, these special adaptations have often been thought to comprise the whole of natural theology. There are far too many of them, it is said, for chance to be the root of them. Therefore they must have been designed, and since neither men, nor spiders nor other scheming beings have designed them, it is simplest to ascribe the plan to deity. As Paley said, we ought to believe in "prospective contrivance".

This pious argument is very familiar. I shall quote from a strenuous supporter in the hey-day of the argument, namely from Henry More in his *Antidote against Atheism*:—

> I demand therefore concerning the cock why he has spurs at all or, having them, how they come to be so fittingly placed. For he might have had none, or so misplaced that they had been utterly useless, and so his pleasure in fighting had been to no purpose. . . . Thus fittingly does Nature gratify all creatures with accommodations suitable to their temper, and nothing is in vain. Nor are we to cavil at the red puggered attire of the turkey, and the long excrescency that hangs down over his bill, when he swells with pride and anger; for it may be a receptacle for his heated blood, that has free recourse to his head; or he may please himself in it, as the rude Indians, whose jewels hang dangling at their noses. And if the bird be pleasured, we are not to be displeased, being always mindful that creatures are made to enjoy themselves as well

as to serve us; and it is a gross piece of ignorance and rusticity to think otherwise.[1]

And again:—

I will rather insist upon such things as are easy and intelligible even to idiots, who if they can but tell the joints of their hands or know the use of their teeth, they may easily discover it was counsel, not chance that created them. For why have we three joints in our legs and arms, as also in our fingers, but that it was much better than having but two, or four? And why are our fore-teeth sharp like chisels to cut, but our inward teeth broad to grind, but that this is more exquisite than having them all sharp or all broad, or the fore-teeth broad and the others sharp? But we might have made a hard shift to have lived through in that worser condition. Again, why are the teeth so luckily placed, or rather, why are there not teeth in other bones as well as in the jaw-bones? for they might have been as capable as these. But the reason is, Nothing is done foolishly or in vain; that is, there is a divine Providence that orders all things. . . . I might add to these, that Nature has made the hindmost parts of our body which we sit upon most fleshy, as providing for our ease and making us a natural cushion. . . . And lastly she has made all the bones devoid of sense, because they were to bear the weight of themselves and of the whole body. And therefore if they had had sense, our life had been painful continually and dolorous. And what she has done for us, she has done proportionably in the contrivance of all other creatures; so that it is manifest that a divine Providence strikes through all things.[2]

Here the counter-argument from dysteleology is formidable. If creatures were made to enjoy themselves, there is a good deal of ignorance and rusticity in supposing that they always succeeded; and they all die, with the consequence that their lives may be plausibly represented as a long fool's errand to the grave. Let us waive the point for the moment, however, and agree that particular teleology in Nature is indisputable and even farther-flung than More's illustrations would indicate.

The essential argument with which we are now concerned

[1] p. 73.   [2] p. 81.

The Argument from Design

is the contention that adaptation can have only one intelligible explanation, namely, design. Mere coincidence on such a scale being absurd, we are told that every particulate adaptation, since it looks *as if* it were planned must actually have been planned, if not by the cock, the turkey and the idiot, then by some other schemer.

In the last lecture I disputed this inference. The particulate adaptations in Nature, as well as the mal-adaptations in nature, sometimes result from conscious planning and sometimes do not. Descriptively we have apparently planned and apparently unplanned teleology and dysteleology. There is no reason at all for holding that the former has secretly swallowed the latter. If we insist upon amalgamating them there are as good reasons for holding that unplanned teleology is predominant as for holding the contrary. Indeed, it may not be wholly fanciful to suppose with William Morris in *Sigurd the Volsung* that both growth and manufacture are late-born patterns:—

Then we fell to the working of metal, and the deeps of the earth would know
And we dealt with venom and leechcraft, and we fashioned spear and bow
And we set the ribs to the oak-keel, and looked on the landless sea:
And the world began to be suchlike as the Gods would have it to be.

. . . .

Belike no fixèd semblance we had in the days of old
Till the Gods were waxen busy, and all things their form must take
That knew of good and evil, and longed to gather and make.[1]

I should like to renew my former argument. If, interrogating experience, we ask what can be done by taking thought, the answer is: "Certain (not all) of our subsequent thoughts may

---

[1] Cf. Laurence Housman:
"Father eternal, ruler of creation,
Spirit of Life, which moved ere form was made."

Theism and Cosmology

be influenced or produced by taking thought. Certain muscular movements may be similarly induced (but not growth or metabolism except indirectly)." Granting, then, that experience testifies to the reality of idead teleology in some instances, we have as good (and indeed the same) grounds for holding that experience accredits the reality of unidead teleology. It does so in ourselves when we grow, and regulate our body-temperature, and secrete what we need; it does so in plants and in other organisms, and in the natural election of inorganic nature. In our own case we can, it is true, purposely affect growth and metabolism by paying attention to diet, by taking certain medicines, by refraining from taking others; and so on. That is not an objection to what I have been saying. We can also magnify the effects of our muscular movements by constructing machines that produce certain planned results. Such machines do not often turn up in nature without conscious artifice. Organisms do turn up without any such artifice. That difference is very important for our practical purposes, but it has no theoretical bearing upon the clear outlines of the difference between conscious purpose and an unidead teleological pattern.

If anyone chooses to say that organisms are natural machines, his metaphor may be permitted or at least indulged. Both machines and organisms are teleological entities. If such a philosopher, however, goes on to maintain that so-called "natural machines" must really be artificial because all machines are artificial, he has forgotten the empirical basis of his own theory; for his assertion was that the particulate teleology of organisms was *not* artificial and therefore was sharply contrasted with the particulate teleology of machines. Indeed, he himself is perched upon a rocking-stone of argumentative instability. He has no empirical evidence for the conclusion that idead particulate teleology is possible except in alliance with a living body, his own or some other animal's. The fact narrowly restricts his own very limited powers in the idead way. Yet he cheerfully asserts, professing to have the backing

The Argument from Design

of experience, that the unidead sort of teleology is wholly superficial, and, in the end, reducible without remainder to teleology of the idead kind.

Plainly his conclusion should be that idead and unidead teleology both occur in nature. If the one is accepted as ultimate fact, the other also should be accepted; and that, I submit, is the terminus of the empirical argument so far as particulate natural teleology is concerned. The existence of Natural Kinds—chemical elements, the metals, the species of animals and of plants is evidence of unidead teleological self-maintenance and cohesion. The principle is not seriously affected if it can be shown that the inorganic elements may be transmuted, and that, in the organic realm, frogs may develop into super-frogs. We are not talking about immutable and immortal teleology either in conscious planning or in unidead patterns. On the other hand there is also the teleology of ideas and of conscious expectations. *Such* planning and scheming has led to the making of cities, and universities, and opium dens, and cathedrals, and music halls.

Let it be replied: "Very well. Have it your own way; but you must admit that these unidead teleological entities (as you call them) *might* have been constructed by a super-human designer, although no mere man, or beaver, or other constructive animal is capable of the feat." The answer is, firstly, that there is no empirical evidence that idead teleology is *ever* the cause of the existence of chemical elements, or of a frog, or of a daisy, and secondly that if we leave experience behind and indulge in untrammelled speculation, it is just as likely that these things were begotten and not made, as that they were made but not begotten. They might be the spawn of an over-worldly fungus as stupid as any toadstool. Such a reply may not disprove the suggestion to which it is opposed. What it does show is that the freedom of such speculations is indistinguishable from mere licence.

Attacking the problem from a near but altered standpoint we may ask whether the notion of planned teleology would

Theism and Cosmology

be at all plausible except on analogy with *our own* planned teleology. Such teleology (as we have seen) is limited, in our own case, to a certain direction and control of subsequent thoughts of our own, and to certain of our own movements, called "voluntary", by means of which we may combine bits of copper and steel into railway-engines, and may also communicate with our fellows by speech and by gesture. Let us then apply this analogy to the theory of a supernatural designer or college of designers.

If the universe consisted entirely of thoughts, we should be dealing with a type of idealism that I am excluding provisionally from this first series of lectures. I may remark, however, that if idealism were the truth, our present problem would re-appear under another set of names. If a railway-engine were a "thought", I could not produce that "thought" (i.e what we call a "real" engine) after the same fashion as I can (perhaps) produce the thoughts that guide my pen and my vocal chords in these lectures. Again, I could not communicate with my hearers except by means of "thoughts" of the first kind, i.e. by means of what is commonly called a physical medium. It is a complete fallacy to suppose that the theory that material and immaterial are both "thoughts" abolishes the plain distinction between material "thoughts" or "realities" and immaterial "thoughts" or "realities". In accordance with my plan, however, I must, for the present, abandon the discussion of philosophical idealism.

Let us return, then, to the provisional realism of this first course of lectures. Is the universe the body of an over-worldly *anima mundi*? If it were, the analogy would be strained. The universe does not seem to resemble an organism (even if it be highly organized). It has still less resemblance to the sort of organism that has a nervous system, and our empirical evidence regarding voluntary movement is limited to *such* organisms. The more important point, however, is that no known organism, our own or any other, is kept alive and healthy, in the primary sense, by scheming and volition. Its

The Argument from Design

deepest vital teleology is unidead. Consequently the conception of an *anima mundi*, so far from suggesting the primacy of teleology of the planned kind, suggests precisely the opposite.

The most usual form of the argument from design asserts that the universe is very like a machine, and not at all like a huge living body. It also asserts that machines are designed, and proceeds to discover that the analogy is felicitous. In this respect, we are told, God has been rather generous to the human race. He has supplied surprisingly good evidence to pious eyes despite their condition of twilight and of mediocrity. Indeed, we should not speak, like a late distinguished theologian, of "the faint and ambiguous suggestions of God which are all that the sinner can detect in creation".[1] Even a sinner, some would say, can see the need for a Celestial Engineer.

Nothing, we think, is more orderly than astronomy. There is no better clock than the clock of the stars; and Professors Milne and Dirac think[2] that a radio-active or atomic clock may be just as good. Men have asked "Of what use are the stars?"; but navigating men have used them to some purpose. We do not believe any longer that the heavenly bodies are composed of different, diviner material than the stuff beneath the moon, but we still believe that they tread so orderly a measure that their harmony is almost literally music. Our telescopes, it is true, do not exhaust the astronomical universe, and some of the galaxies beyond the Milky Way may be less orderly than our own galaxy, but there is no reason for supposing that the laws of motion are local cis-galactic habits and nothing more. Assuming, then, that suns and planets are simply material bodies, not living beings or gods,[3] assuming even that they are mostly holes, they may more reasonably be compared to a splendid machine than to anything else we can descry. It is therefore argued that since machines do not turn up in the world unless they are designed, the world-

[1] H. R. Mackintosh, *Types of Modern Theology*.
[2] *Nature*, February 1937.   [3] As in Greek astropsychics.

machine itself cannot be supposed to be something that just turns up. So it must have an over-worldly designer.

Let us try to discriminate some of the principal things that may be intended by this argument.

Firstly, the argument from design, like all theistic arguments, is meant to reach the ultimate. We do not ask who designed the designer, but regard the designer as, in Hobbes's phrase, "ingenerable". That may be legitimate if the premiss that every machine is designed by a mind, and not by a machine, is accepted. Such a premiss does not state that the designer (who is not a machine) is designed, although it might be argued on other grounds that everything must have a cause and so that the designer is not ingenerable. (There are no known designers who *are* ingenerable.)

Secondly, it is plain that the argument implies a designer (or designers) who has (or who have) the power to produce the astronomical machine, and whose purposes may in part be inferred from the characteristics of the machine produced. It does not imply more than this. Therefore it does not justify the inference that the designer is a self-necessitating being, or has the unlimited attributes commonly ascribed to deity by orthodox Christian theologians, or is the sort of being that is called God in the literature of the major historical religions. This is sometimes regarded as a signal advantage, as by Dr. Tennant when he writes: "The empirically-minded theologian ... asks how the world, inclusive of man, is to be explained. He would let the Actual world tell its own story and offer its own suggestions. ... He will thus entertain, at the outset, no such presuppositions as that the Supreme Being, to which the world may point as its principle of explanation, is infinite, perfect, immutable, supra-personal, unqualifiedly omnipotent or omniscient. The attributes to be ascribed to God will be such as empirical facts and their sufficient explanation indicate or require."[1]

In the *third* place we have to observe that the argument is

[1] *Philosophical Theology*, II, 78.

The Argument from Design

no longer particulate, as in Henry More's remarks about the cock and the turkey, but has now become, so to say, oecumenical, and indeed more than oecumenical if that word is understood in its literal sense as applying to the *inhabited* world. It is the whole universe (if it be a whole) that is said to be the machine-like thing that requires a designer or college of designers. Special machine-like adaptations within the great machine are consistent with the theory, and may be on far too vast a scale to be the work of any finite designer empirically known. But the oecumenical and not the particulate standpoint predominates.

Consequently the Argument from Design has to withstand the venerable but active objection that it is illegitimate to argue from what uniformly happens *within* the world, to what must happen *to* it. (We cannot say "uniformly happens to it" since the world is unique.) No one would maintain, for instance, that because every individual Scotsman has had a father and a mother, therefore the Scottish people, or the Corporation of Glasgow, or any other organization of Scotsmen must have had a father and a mother. The traditional answer would be something like this: We discover causes and effects by presence-and-absence methods of varying degrees of refinement. Hence numerous instances are required for making the discovery, but once the discovery has been made it may be extended forthwith to a single new instance. That, we may be told, applies to the present case. The Scottish people is an organization but is not an organism. If it were an organism of the human type we could infer that it had had a father and a mother even if we had no other evidence of their existence. Since it is only an organization and not an organism, that particular inference is prohibited; but there is no such obstacle in the way of the Argument from Design. If the astronomical universe really is a machine, and if we are taught by experience that machines *must* be designed, there would be no greater measure of unreason in inferring from the evidence of the parts, or of a sufficiency of them, that the whole is designed, than for inferring, say, that the whole is extended or in process because the parts, or a sufficiency of them, are extended and are in process. Parts and whole *may* have the same properties. In the absence of any known reason for discriminating between them in this respect, the inference from the parts to the whole should commend itself to a sober judgment.

Hence we are led to the two major considerations that we may label the fourth and the fifth respectively. The fourth is whether we are really entitled to say that the astronomical universe is a machine, and the fifth is whether, if it were machine-like, its machine-like character might not simply be ultimate. [I may say, parenthetically, that I prefer the term "machine-like" to the term "mechanical". The latter pretends to a precision that it does not possess.]

Take, then, the *fourth* point. I have argued that the mere fact of adaptation and of self-maintenance does not entitle us to say: "Here is a machine that must have been designed". Consider, for instance, the stable chemical elements, or relatively stable substances like sugar or wax, or the Rutherford model of the atom. Even if such teleology is machine-like in certain senses, we dare not say that *these* machine-like things have been designed by any known mind. There is no proper analogy, say, between the Rutherford atom and an Ingersoll watch. Paley justly remarked that if one found a watch on a desert island, one would infer that there had somewhere been a watchmaker.[1] Watches do not turn up like rocks. But atoms do. So Paley should not have inferred that *atoms* were designed.

This leads to the *fifth* point. Why should we deny that the order and regularity in the astronomical universe may simply be ultimate fact? *Something* must be ultimate; if not the world, then its cause; if not its proximate cause, then God. Why should anyone obscure the issue by saying that the orderliness of the astronomical universe cannot have occurred at random? It is certain (I submit) that everything that exists must be

[1] *Evidences*, chap. i.

The Argument from Design

determinate. Therefore the mere fact that the universe is determinate, is just what it is, has no bearing on the issue. It is not a question of "chance" at all. We have no right so much as to say that it *might* have been different. We can only say that we may suppose it to have been different if we choose, and may debate that supposition.

Is the contention that it does not stand to reason that material particles should, as the phrase goes, "obey" the laws of motion? Certainly we cannot say *a priori* that motion must be uniform, or that, if it be uniform, its first law should be in terms of rectilinear and not in terms, say, of circular motion. In a sense, therefore, the laws of motion may just be brutally factual; but why should anyone suggest that it must be possible to get behind such "brutal" facts, that because we cannot see *a priori* why the thing must be so, we are entitled to hold that there is a curtain that can be pulled aside? Is the argument that, granting the uniformity of the laws of motion, there must be a specific collocation of moving particles if the particulate facts of the world are such as we find them to be? *Some* collocation there must be. Why not this one? It is only if the evidences of design are of a very special kind that any such argument would have so much as a tincture of persuasiveness.

I anticipate the retort: "*Of course* it is the special marks of design that are so significant. Isn't that where your lecture began with its remarks about fighting-cocks and turkeys?". In a sense I agree. We have not advanced very far, but if our discussion has been rather like a spiral, the whorls are narrowing.

They have been narrowing, too, in another way. The argument to a designer of the universe, when it is based upon empirical analogies, seems stronger the liker the universe is to a machine. We were therefore prepared to allow that the astronomical framework of the universe was *very* machine-like, and to develop the consequences. For the most part, however, the advocates of an empirical theology base their views upon *all* the evidences of design, including (they say) life and mind. But life is not very machine-like, and minds, to say the least, are not more machine-like than life is. There is no empirical evidence that they were manufactured on purpose (as there is in the case of most machines). No doubt an omnipotent creator *might* have manufactured them; for, *ex hypothesi*, he *might* make anything. They seem, however, to be born and to grow. If it be held that our planet was once incapable of sustaining anything at all similar to the life that we know, and consequently that the maxim *omne vivum e vivo* cannot hold of the entire history of life upon the Earth, the conclusion is that the ultimate origin of terrestrial life may be very mysterious, but scarcely that life must have been manufactured like a machine.

I do not pretend to guess where the boundaries should be drawn in this matter. Many people seem to think that there is no great difficulty in believing that living beings acquire mental properties by an easy and natural progression. What staggers their imagination is the origin of life itself. To others the emergence of logic or of aspirations towards beauty and rectitude in living things is something that must be altogether meta-biological. They are quite prepared to say with Descartes "Deus corpus nostrum fabricavit ut machinam",[1] and to believe that bio-chemistry and bio-physics are complicated departments of chemistry and of physics and are nothing more. We need not take sides very strenuously upon this question. What concerns us here is the simple circumstance that the argument that what is machine-like implies a designer stumbles badly when it proceeds from the astronomical framework of the universe to biological and to psychical teleology. For one of these at least is very unmachine-like.

But let us examine what are supposed to be the special indications of God's designs.

Design is idead teleology. Therefore we are not entitled to argue from what we allege to be special marks of design

[1] *Entretiens avec Burman*, ed. Adam, p. 70.

The Argument from Design

unless we lay claim to a certain appreciation or understanding of the designer's purposes. In our earlier argument, at the first whorl of our spiral, it might have been enough to say: "The world is a machine; we don't know why it was designed, and we needn't guess; but at the lowest rating it must have been designed." The closer argument we are now considering, at a higher whorl of the spiral, is that there are special marks of special design. Many philosophers, of course, would repudiate this contention altogether, perhaps saying piously, as Descartes did in his conversations with Burman, that it would be unworthy of deity to suppose that he had any purposes. *Deo maxime indignum.*[1] That, however, is precisely what would *not* be said by the "empirical" and "unpretentious" theologians who prefer the Argument from Design to any other.

What are the designs of God that, as it is claimed, men may read in the book of nature and in the book of human history? Some would say "His own glory", and although the phrase may suggest a magnified non-natural Roman emperor, it need not now be challenged. In this case the argument relies upon a presumption regarding the good pleasure of the most majestic being conceivable. "For his own glory" God made his universe vast and orderly and beautiful. These, I think, are the principal suggestions regarding the book of nature. It is a splendid volume. The next suggestion concerns life, and as we do not commonly think (although here we may very well be wrong) that mere vitality, without sense or reason, could glorify God in any appreciable degree, we say that the happiness of God's creatures is one of his chief designs. This applies to the cock and the turkey and to some very nasty animals besides. In the main we think of man as God's noblest work; an honest man, a saint, a dreamer, a lover, a poet. Empirical theology, when conceived on these lines, has these divine purposes chiefly before its gaze.

Let us then consider the book of nature, its vastness, its order, its beauty, and examine these special marks *seriatim*.

[1] Ed. Adam, p. 70.

Theism and Cosmology

First, then, vastness. After all, size is relative; but, dropping the point, I cannot see that any cogent argument lies in respect of mere vastness either spatial or temporal. In default of a more elaborate discussion, I may recall your attention to a conversation that took place in Eden and was reported in the eighth book of *Paradise Lost*. Since Adam seemed to be "entering on studious thoughts abstruse", Eve "went forth among her fruits and flowers", not (we are carefully told) because she could not follow the discussion, but because she preferred to talk it over connubially with Adam when Raphael the "affable Archangel" had gone. Adam's puzzle was why the Earth, "a spot, a grain, an atom with the Firmament compared" should be served by so many stars

> merely to officiate light
> Round this opacous earth, this punctual spot
> One day and night, in all their vast survey
> Useless besides.

He could not understand

> How nature, wise and frugal, could commit
> Such disproportions, with superfluous hand
> So many nobler bodies to create.

Raphael replied like a politician. Adam would be wiser not to try to decide between the geo-centric and the helio-centric hypotheses; but, as regards vastness, there was little substance in his perplexities.

> Consider first that great
> Or bright infers not excellence. The Earth
> Though, in comparison of Heaven, so small
> Nor glistering, may of solid good contain
> More plenty than the Sun that barren shines,
> Whose virtue on itself works no effect,
> But in the fruitful Earth; there first received,
> Her beams, unactive else, their vigour find.

The Argument from Design

> Yet not to Earth are these bright luminaries
> Officious, but to thee, Earth's habitant.
> And for the Heaven's wide circuit, let it speak
> The Maker's high magnificence.

It is to be feared that the Archangel's last remark, doubtless because he was arguing *ad hominem*, is not fully consistent with its predecessors; but he said enough effectively to puncture Adam's scruples regarding size.

Consider, next, order. It was the perfection of the orderly motion of the heavenly bodies that led so many Greek philosophers to regard these orbs as more divine than the denizens of Olympus or any other godlings that poets or rustics had fancied. In the seventeenth century, when the new physics triumphed, Christian philosophers rallied their forces, and found the new regulations of cosmic order even more encouraging to theists than the old. They had to allow that God performed occasional miracles, but the bolder among them, such as Malebranche, averred that God performed very few. Locke, Malebranche's contemporary, found the ultimate proof of the reasonableness of Christianity in miracles; Malebranche did all he could to belittle miracles. According to Locke the miracles were "oracles of God and unquestionable verity. For the miracles [Christ] did were so ordered by the divine providence and wisdom that they never were nor could be denied by any of the enemies or opposers of Christianity. . . . The enemies of Christianity have never dared to deny them, no, not Julian himself."[1] Malebranche, on the contrary, thought it unworthy of deity that he should interpose here and there, like a clumsy watchmaker correcting a faulty timepiece by hand. The love of order was the supreme divine characteristic, and although the divine love of order (according to Malebranche) involved the dolorous consequence that somewhere between 95 and 98 per cent of the human species should suffer

---

[1] *Reasonableness of Christianity.* Locke's *Works*, tenth edition, pp. 135 and 138.

Theism and Cosmology

eternal torment, this frightful consequence, whose horror Malebranche never attempted to disguise, was as nothing (he said) in comparison with the glories of the "order" that was supposed to imply it. From harmony, from universal harmony, this universal frame began. It drowned the groans of the damned in its diapason. Only a miracle could save them, and a miracle was a deeper discord than their unquenchable misery.

To such lengths humane and gifted men felt compelled to go in defence of order as such; but surely their argument was very weak. If deity ordained everything for the best, the events in the universe would exhibit the fullest perfection whether God made them conform to a few simple laws or constantly interposed like a finicking artist who will never let his work alone. In short, it is a special sort of order that is tacitly understood in these arguments, the sort of order that is relatively simple for human beings to comprehend, not merely because the laws themselves are simple, but chiefly because they show their simplicity in certain notable and manifest recurrences that wayfaring men, although stupid, cannot permanently miss. This is the "order" of simple austere architecture, and commentators should have a certain sympathy with Spinoza when he regarded order as an aesthetic attribute, and not as a feature of demonstration *more geometrico*.[1] In so far as the universe is God's geometry it shows, no doubt, a geometrical order; but Spinoza was thinking of the glory and beauty of "order", not of the simple fact that divine geometry had to be logically consecutive.

Both in the alleged majesty of vastness therefore (as in the "affable" Archangel's lecture) and in the interpretation put upon "order", we seem to be gently introduced into aesthetic considerations, and it may be asked whether beauty be not itself the strongest link in the argument from design. According to Samuel Alexander, "It is (to drop into flagrant metaphysics) much truer to think of God as an artist than as a mathematician

[1] *Ethics*, I, Appendix.

The Argument from Design

as some have proposed. For the forms of art are not imposed as would be the form of a mathematician-God upon a foreign and recalcitrant material, but the material of the creator is itself moulded into the forms he gives it."[1] This comment may be hard upon a geometrizing God, for it is needless to conjecture that space is a foreign and recalcitrant "receptacle" to him, and it would not be true at all of a deity whose tastes were algebraic. Nevertheless if God designed as a painter, a musician, a sculptor or a poet designs, if the Logos were a noble period of divine rhetoric, design might be plainer than if he were essentially a world-carpenter, a world-potter, a world-engineer. Kant devoted the first and the most interesting part of his third great *Critique* to a form of this problem. Schelling preached an aesthetic theology. The theory, therefore, has high historical sanction. As theologians, φιλοκαλοῦμεν καὶ φιλοσοφοῦμεν.

At its first look, the aesthetic argument from design may be stated without any reference to man or to his mind. For natural beauty is sometimes supposed to exist where there is no human, or animal, or angelic beholder. In this form the aesthetic argument would come nearer to the spirit of the usual sort of cosmology (and to the present whorl of our spiral) than in any other form. It is difficult to see, however, how an *inhuman* aesthetic argument from design could be expected to produce much conviction, although some modern philosophers find it "well-nigh irresistible".[2] Personally, I should like to think that things may be beautiful in the simple sense I have mentioned above; but I should not expect anyone who shared such views with me to allow, further, that natural beauty must be designed, that is to say, must be a work of art. No such inference could be drawn from empirical observation. Moreover, I have to allow that comparatively few philosophers would so much as entertain the idea that things, in any simple sense, can be intrinsically beautiful. Their so-called beauty,

---

[1] *Aristotelian Society Proceedings*, 1936–37, p. 133.
[2] G. Dawes Hicks, *The Philosophical Basis of Theism*, p. 230.

according to the vast majority of writers, is a way of saying that they evoke an appropriate admiration in us—and not a very good way of saying it. Some indeed would say with Hegel [1] that Nature is too stupid to be beautiful.

This last is a bad argument. It confuses beauty with artistry and it is opposed to experience; but true opinions may be supported by bad arguments; and the naïve and heterodox view that I like, namely that natural objects may be beautiful in themselves, can hardly be regarded, with any confidence, as even a slender pillar of theism. It is difficult to be convinced, on these lines, that Nature is what Sir Thomas Browne said it was, "the art of God".

Most beauty is sensuous. If a good realist maintained that colours, sounds and other sensible properties are genuine features of the world and could exist although nobody saw them or heard them, I should salute him and wish him all power to his larynx. All the same, I am sure he could be forced to admit that what different men see and hear varies with the state of their sense-organs and with other individual conditions. The consequence would be that any one man makes a different sense-selection from any other man. Even if these individual differences did not usually make any great difference to the cruder sweep of sense-apprehension, they would make a very great difference to the finer nuances of aesthetic appreciation. If birds have a sense of beauty—and this has been held—we can scarcely believe that what they see and hear has more than a general resemblance to what we see and hear. If angels can listen to the music of the spheres—and there is good authority for holding that they cannot since they are mere intelligences —we have no right to suppose that their sense-selections are the same as our own. It is also heretical (although it might be true) that God, who is impassible according to many theologians, here resembles the men that are made in his image. In short it is very difficult to dispute the presence of a concealed anthropomorphism in these aesthetic arguments, or to

[1] See E. F. Carritt, *The Theory of Beauty*, p. 36 n.

The Argument from Design

avoid saying with Francis Bacon that beauty has to do with "the shows of things fashioned to man's desire".

Natural beauty, then, is evidence, primarily, of a certain adequation or conformity between mind and its milieu. The nature of the adequation or conformity is, of course, much more disputable. Beauty is subtler than the simple evocation of human delight, and it may, as Kant thought, have its metaphysical explanation in the harmony between human faculty and its world, not conceptually, but in a more intangible atmosphere of agreement. The finer shades of this delicate topic, however, do not (I think) concern us here. If they did, their basis would be, not divine artistry, but a union deeply interfused between man and sensuous nature. A lovely universe may be a stronger argument in favour of pantheism than a friendly or than a reliable universe, and it might be held, without internal contradiction, that although radiant beauty is very rare, "there is a certain level [of beauty] below which nature never falls".[1]

The last of these statements, moreover, seems to be very insecure. I quote from the impression that the Far North, in summer, made upon Lord Tweedsmuir:—

> Aldous Huxley somewhere asks whether the Wordsworthian peace and healing-power of Nature could be found in a lush tropical forest where Nature is cruel and fantastic. I think there would be the same difficulty in the North.
>
> Wordsworthian Nature should have a certain reticence, cleanness and simplicity, and be built more or less on the human scale. The North is altogether beyond the human scale, and it has no cleanness and simplicity. I expected bleak moors and cold ice-grey waters. Instead I found a kind of coarse lushness—immense rivers, pouring billions of dirty gallons to the ocean, too much coarse vegetation, an infinity of mud, and everywhere a superfluity of obscene insect life.
>
> The impression was not of a Nature beautiful and austere, but of a Nature as coarsely exuberant as a tropical forest. The delta

[1] G. Dawes Hicks, *op. cit.*, p. 228.

Theism and Cosmology

below Aklavik has vegetation as rank as the delta of the Zambesi. The streams are as thick with mud as the Limpopo. I took a "Wordsworth" with me and found him perfectly unsuitable. My impression was of a monstrous superfluity of rank vegetation and mud. It was only in the miraculous sunsets that any grace or magic entered into the landscape.

. . . .

The Arctic coast was different. There among clear half-lights and half-tones one had a wonderful feeling of space and peace. Great Bear Lake, with its immeasurable deep blue waters and its rocky shores, like the West Highlands, gave one the proper feeling of the clean antiseptic North. But I think the most wonderful impression I had was flying over the Barrens on a cloudy day. The cloud shadows in these infinite plains in constant motion made a beautiful fantastic world. It was all out of scale with humanity; but it is a good thing now and then if you manage to realise that the world was not created on your own scale. It sharpens the adventure of living.[1]

These are confident opinions, and not unprejudiced, despite the writer's air of judicial balance. By various routes, however, we arrive at the conclusion that when God's special purposes are said to be discernible in the adaptations and adaptability of the world, there is usually a covert reference either to mankind or to the sort of human ideals "antiseptic" or "obscene" that a man prizes or abhors when he reflects upon his own very human experience.

In other words, we are moving towards a discussion of God's providence in the special form of his tender care for mankind. That is an aspect of the sort of question we are reserving for the second course of these lectures. A certain precipitancy, however, can scarcely be avoided now, although I shall try to restrict it to the topic immediately confronting us, namely design.

The fitness of the earth's crust to human life and happiness, and the fitness of man to exploit the earth's crust, are the most

[1] "Canada's Far North", contributed to the *Sunday Times*, 1937.

The Argument from Design

obvious themes of the kind, together with the moral argument that the earth's crust seems to some to be a vale of soul-making. The intelligibility of the universe, that is to say, the relatively high degree of its intelligibility to clever men, is another similar argument. Clever men might have been much scarcer than they are, or might have had a much harder job. These lines are capable of bearing further traffic. If the earth's crust is peculiarly adapted to human happiness it must also, by parity of reasoning, have a certain marked adaptation to the happiness of fighting-cocks and of turkeys, tunny-fish and man-eating tigers. Other planets, too, may be peculiarly adapted to beings who can think, suffer, admire and adore, even if these beings, in other ways, are very unlike mankind and man's furred and feathered terrestrial companions.

Here we have firstly to examine a problem in metaphysical ecology. Does the fitness of the environment to living, joyful, intelligent and righteous beings compel the inference that the environment must have been planned for that purpose?

Let us consider man's estate. Are there convincing indications that Nature as a whole must have been designed as a favourable habitat for human convenience, human hope, human morals and human science? Is it inconceivable that the fitness of the earth's crust for these affairs, as evidenced by humanity's actual success, or by its reasonable confidence in future success, should be, as it were, a cosmic by-product requiring no special explanation in terms of design? Are cosmologists likely to echo Irenaeus in a much-quoted passage: "God arranged everything from the first with a view to the perfection of man, in order to deify him, and reveal His own dispensation"?[1]

Take the human intellect. We hold (not indeed universally, but generally, firmly, and, as I think, justifiably) that man's intellectual discoveries are not, in principle, discoveries of what is merely human. If two and two make four, the fact is not that human two-and-two's make human fours, but that

[1] *Adv. Haer.*, V, 36.

Theism and Cosmology

two and two really do make four. It takes an intellect to discover logical implications, but the fact so discovered is no more a human fact than the sun, if there really be one, is a human sun. Accordingly, if the fitness of the environment to the human intellect means *only* that there is something in that environment that human intellects can discover, it may be unreasonable to expect that a proof of the existence of a designer can be extracted from the circumstance.

Therefore the argument usually takes another form. Unless Nature exhibited visible and experimentally controllable regularities of a comparatively simple type, we are told, our intellects would be helpless except, perhaps, for a few metaphysical tags like "ens est res", "res est aliquid", or for the "tautologies" of strict logistical positivism. What has to be explained is why the world is such a promising laboratory. I cannot think, however, that the argument in this altered form is at all cogent. How do we know that the terrestrial laboratory is so very promising? It is commonly held to apply to phenomena and not to things in themselves, to tell us not what things are but how in the past they have been sensibly experienced. I allow that realists strenuously dispute these latter arguments, but even if the arguments were subject to a heavy realistic discount they could not prove that the universe was designed to give scientists a comfortable innings.

Take human happiness and human suffering, a theme more narrowly human, since if X is pleasing to man it need not be pleasing to any other creature. Have we to say that the natural "scheme of things entire", must be a "scheme" expressly designed for the zest and contentment of human beings?

It should be conceded, I think, that many of the current objections to this conclusion are rather weak. It is said, for instance, that the earth's crust is a very small mouse for a labouring cosmos to produce, but, as we have seen, we have no yard-stick for measuring the appropriate ratio between the

The Argument from Design

value of a result and the magnitude of the means for securing it. It is quite illogical to infer the insignificance of man from the collapse of geo-centric astronomy, and it is false in fact to suppose that geo-centrism did puff men up. On the contrary, many of the mediaevals deplored the supposed fact that they were as far as possible from the highest heaven, the farthest removed from its radiant but unequal power. Again, if it be complained that man's happiness has been a long time in coming, we have no standard for affirming that the time was unreasonably long. The argument might be a serious objection to an omnipotent designer (who would need no time at all) but it need not affright an empirical theist who is content with a theory of a limited deity. Once again, the argument that man and all his works must perish from world-entropy in an implacable if distant future seems stronger but may be upset by a newer scientific discovery.

Nevertheless the argument from dysteleology seems to be disconcertingly formidable. We are now talking, be it remembered, as cosmologists, and are asking, firstly, whether it is reasonable to believe that the earth is by nature man's garden and his pleasance; secondly, whether a critical cosmologist should infer that the whole of the universe is a means to man's felicity, or, more widely, to the felicity of sentient beings on the earth and a few other planets. Is the first of these premisses at all secure?

I am not disputing the philosophy either of a temperate optimism or of a guarded meliorism regarding human felicity. Men, on the whole, have wrung a good deal of contentment out of their lives. Human life, on the whole, seems to have been pleasant, not merely tolerable. Perhaps, as Maecenas said, *Vita dum superest, bene est*. Death, in the main, need not be cruel, and may usually come neither too soon nor too late. Even if the principal source of human happiness were the illusions of hope rather than the experience of joy, that illusion would itself be comforting and sustaining. In that sense, Leibniz may have been right when he said that a happy

Theism and Cosmology

folly is better than a bitter prudence.[1] That, speaking roughly, is what I mean by a "temperate optimism" in this matter of human felicity. Similarly I mean by a "guarded meliorism" that there is a reasonable prospect of greater human happiness both in the near and in the remoter future. That is not certain and the promise of it may easily be destroyed; but the diminution of poverty, disease, ignorance and the more disastrous sort of folly is something more than a glowing dream.

Let that be granted, and let it also be granted that the studiously neutral attitude of natural science in this matter need not be more than a prudent policy. Making these admissions, or even greater ones, are we entitled to affirm that the favourable balance towards felicity in human existence is so marked as to require a special explanation; that the incidence of these abundant means towards human happiness resembles manna from heaven rather than the hedonic neutrality of the snow and the rain? To me at least it appears that we are not so entitled. Our data are far too meagre for such conjectures concerning the probabilities. The debate is vain, because the terms of it are hopelessly indeterminate. There is much happiness and much misery. An overplus of happiness, supposing it to exist, would not seem to require any special explanation in terms of design.

Lastly, in this connection, let us turn to the moral argument from design in the form of it that concerns cosmologists. Is the earth's crust so peculiarly well adapted to righteous human conduct as to imply its design for that end?

On certain assumptions this argument is not independent of the last. According to utilitarian moralists of the traditional school, the rightness of moral action is a function of its skill in procuring happiness. Many of these moralists were secularists. They assumed only that happiness could be procured, and that much pain was avoidable. Some of them, however, were theological hedonistic utilitarians. Their thesis was that God had designed the happiness of his creatures, and had taught

[1] *Philosophical Writings* (Everyman), p. 248.

The Argument from Design

men the major principles of happiness-getting, including, most signally, moral rules. The fitness of the environment, on these assumptions, would be a divinely designed precondition of social and of individual prudence.

A more recent school of utilitarians, commonly called "ideal" utilitarians, deny that welfare is simply an affair of happiness and of the avoidance of pain. In their view there is an ultimate plurality of intrinsic goods, happiness, intellectual insight, the experience of beauty, and, it may be, many others. A theological ideal utilitarianism, conceived on these principles, would raise the same general questions as we have already considered, namely whether God must have designed the world as a school for intellectual insight, aesthetic education and the like.

Certain of the ideal utilitarians also hold, however, that many moral qualities, such as candour, compassion and purity have an intrinsic dignity of their own, and so belong to the ultimate plurality of intrinsic goods. This theory may not be quite strictly utilitarian, and it may have to walk very delicately if it is not to count some of its goods twice over. Supposing, however, that it could be consistently maintained, a fresh point would undoubtedly be raised, namely the fitness of the environment to the growth of these special qualities; and that is one of the meanings that could be given to the philosophy that regards the world, Keats-wise, as a vale of soul-making.

The same would be true of the ethical school that, in the past, has been most persistently opposed to any form of utilitarianism, that is to say, to the school of deontologists in morals. According to them, right is right and there's an end on't. The property of rightness pertains to human actions and can be seen to do so by rational insight. Such insight also reveals that rightness is not a means to anything else, unless incidentally and in a way that is morally negl'gible. According to all such moralists, therefore, the fitness of the environment for the education in the performance of duty is totally distinct from the fitness of that environment for procuring happiness,

Theism and Cosmology

avoiding misery, or, in general, from promoting any non-moral type of welfare.

On the whole, the tendency of most of the recent expositions of deontology in ethics has been to keep aloof from the metaphysics and from the ecology of the subject. This attitude has a certain support in the logic of the conception. Duty according to that logic is duty in its own right. The only metaphysical or ecological requirement is that duty can be performed. On the other hand, even deontologists would admit the importance of moral training and of an environment rich in moral opportunities; and the dignity-school of moralists (whether or not they are ideal utilitarians) would admit the relevance of such questions rather eagerly.

I find it difficult to believe that cosmology has any clear message about these matters. We may certainly maintain that the school of nature is hard, but not oppressively so; that the apparent ethical neutrality of rain and tempest, sea and earthquake, is salutary on the whole; and the like. Some moralists believe that the school would be too hard were it not that the helpfulness denied to mere nature is supplied, counteractively, by divine grace. If so the argument would go beyond cosmology; for grace in the relevant sense is adjudged supernatural. In general, however, the argument traverses the ground we have already attempted to explore. The opportunity to relieve suffering depends on the prevalence of suffering. The opportunity to preserve and not to squander life and thought depends upon the chances of longevity and of sustained thinking. If we lived in heaven we should rejoice that sin and sorrow had been done away, and that we did not have to fight them. Moral ecology, in short, is like the fabled bird of Paradise. It has no feet. The subject is congenial to apologetics; but apologetics is a pursuit in which the goalkeeper is also the referee.

As we saw, these arguments concerning the suitability of the earth's crust to human purposes and ideals have an analogous application to those of the animals that have some

The Argument from Design

similarity to humanity. Thus if man's "plenty" be defined as Mr. Hogben defines it, namely as "the excess of free energy over the collective calorie debt of human effort applied to securing the needs which all human beings share",[1] other animals may have a surplus when they have paid their calorie debts, although their unscientific expenditure of energy is likely to be rather wasteful.

In the main we think of this question in terms of animal contentment and of animal suffering; but the evidence, as in the case of human contentment and suffering, is indeterminate. It is likely enough that when biological nature is depicted as a field of carnage, the sufferings of the combatants are enormously less than the death-struggles of educated and imaginative nations impressed by the million into a scene of thundering metallic slaughter. The unimaginative combatants in desert, field and forest may suffer comparatively little either in life or in death. There is no doubt, however, that they do suffer, and that the credit-balance of their joys, if it exists, is a matter of unverifiable speculation.

It is also permissible to opine, in a speculative way, that planets other than our own, perhaps in our own solar system, may be inhabited by intelligent and sensitive beings, although these Martians (or others) could not be very like us and our animal companions. Other solar systems may have planets with a still more suitable ecology.

Such speculations have usually been repugnant to those who would regard cosmology as in its essence a redemptive drama, where the Cross is the culmination of the most moving story in human literature. The story would be cheapened if it were not unique. On the other hand these speculative possibilities would certainly be a solace to those cosmologists who boggle at the idea that the whole natural universe is only the stage for an all too human drama. If, again, the truth be that all natural existence has life-like properties locally contracted into plants and beasts and men, it seems unlikely that

[1] *The Retreat from Reason*, p. 71.

Theism and Cosmology

nothing similar to what we find upon the earth should occur anywhere else.

Wider speculations of this order are relevant to the general problem of the place of teleology in Nature, but have only the loosest relation to *planned* teleology or design. For the same reason it is most unlikely that these discussions of a planned ecology would be strengthened, in a sense favourable to the argument from design, if we considered, not the fitness of the environment to the users, but the origin of the users themselves. The argument from design in its sanest, most circumspect and most impressive form is an empirical argument which infers a designer on empirical grounds, and declines, of set purpose, to enlarge its inferences beyond the point where a designing cause appears to be the most reasonable hypothesis for explaining the facts. My contention has been that teleology and planned teleology are not the same thing, and that, empirically, the users (and particularly life and mind) appear to arise in accordance with the *unplanned* and not with the planned species of teleology. If they are planned, as the size of a family may be planned, the planning presupposes and is dependent on the unplanned teleology of conception and of gestation. Similarly, although minds may act in accordance with a plan, the origin of a mind, so far as empirical evidence goes, depends upon unplanned biological teleology.

Indeed the existence of the user, if the user be an organism or a minding organism, is an obstacle in the way of the argument from design and not a support to that argument. If with Plato in his *Laws* we take our stand upon the origination of motion, then *every* self-moving being, man or beetle, would, in terms of the argument, be a mind or a God, that is to say, something self-moving and not other-moved. Its origin would therefore be exempt by definition from the designs of a Maker. If we take our stand upon "first causes", and hold that the voluntary actions of men are "free" or "first" causes, then we contradict ourselves if we go on to affirm, for better measure, that these "first" causes are designed by God. If God

The Argument from Design

designed them they would be *caused*, not uncaused causes. So here. There cannot be health in the attempt to show that unplanned teleology should be explained by planned teleology.

The same would be true if the whole pattern of reality were teleological. It need not on that account alone be *designed*; and if the pattern, so far as we could see, belonged to teleology of the unplanned type, then, so far as we can see, the fact would be hostile to the argument from design and could not be its ally.

Thus we may be told that Nature develops the sensitive nature of fishes of birds and of beasts. She develops the reminiscent nature of the elephant that never forgets. She develops her Aristotle, her Galileo, her Newton, her Mendel. She may have developed, or she may be about to develop, something still better than the mind of a great man. That (it may be said) is a part of her teleological pattern. She declines to be sundered, and admits no gaping fissures between minds and mindless things, or between vitality and utter death. She also declines to banish life or mind into a Limbo of non-natural things. She is biological and also meta-biological. She is psychic and, peradventure, meta-psychic.

These statements may be figurative and rhetorical; but I am not disputing the substance of them. I am also not disputing the pantheistic theory that a godless "Nature" is a vicious abstraction. What I have been discussing is the argument from design regarded primarily as an empirical argument and applied philosophically to cosmology. I am asking whether these or any other such statements, stripped of rhetoric and of metaphor (or, for that matter, bedizened with rhetoric) do actually describe design on the part of any being. It seems to me that they do not. I would as soon speak, with a late distinguished British statesman, of "the mind of events".[1] I have the greatest respect for teleological argument in this matter, but little or none for the Argument from Design.

There is, quite certainly, teleology *within* nature and there

[1] Grey of Fallodon, *Twenty-Five Years*. Grey used the phrase apologetically.

Theism and Cosmology

may be teleology *of* nature. The latter may be the deiformity of nature, the former may be a sub-pattern echoing the major pattern. I could not dispute these things if I wanted to dispute them; and I do not want to dispute them. On the other hand the Argument from Design seems to me to be quite fantastically weak, and theists, in my opinion, would be wise if they abandoned it, putting their faith in the deiformity of the teleology in things, and not upon their *planned* deiformity. If there had been no confusion between teleology and planned teleology I cannot believe that the Argument from Design[1] would have found very many adherents.

If I were asked whether the Argument from Teleology gives stronger support to theism than the Argument from Design, I should say that the Argument from Teleology is much the better based, but that its theistic tendencies are highly speculative. The theist's problem rests not simply upon the presence of teleology in nature—for that is apparent—but upon the kind of teleology that is present. Since unidead teleology may be sub-vegetative, it would not seem that teleology, as such, provides a proof of a spiritual or of a super-spiritual teleological co-adaptation. For any such proof to prevail intermediate propositions would have to be discovered, and all of them would require a much more extensive exploration of mind, spirit and value than are appropriate to this first course of lectures.

---

[1] A more accurate name for the Argument would be "The Argument *to* Design" or "*to* a Designer."

X

Examination of Cosmological Theism

In his *Opus Posthumum*, Kant, borrowing the term, perhaps, from the title of a book by Huyghens, declared that reason was *Cosmotheorus*, at once the original spring (Urheber) and the indwelling spirit (Inhaber) of the cosmos. I have not tried to speak, like Zeno the Stoic, in the name of Reason, but still I have been dealing, for all my littleness, with things that are very big. Provisionally realistic, these lectures have avoided ontological idealism for the time being, and have not discussed whether all reality must be mind-constituted. It is only in a very modest way that their philosophy could suggest a "spectator of all existence". Indeed, they did not expressly affirm that existence is a single comprehensive and comprehensible entity, although they had sometimes to use expressions like "the cosmos", the "world", the "whole". Nevertheless our theme itself compels us to be *cosmotheori*, and now that we have reached the concluding lecture in the cosmological part of our enterprise, it seems necessary to review the position. The present lecture will be in part a recapitulation, but I hope also to include in it a freer discussion of the general question.

I have argued that no theism, scientific or poetical, hypothetical or professedly demonstrative, empirical or *a priori*, is worth calling by that name unless it is, in intention, ultimate, and unless it holds that deity operates on a scale that is at least as wide as the cosmos. I take both these points to be obvious from the very meaning of the enquiry. Since pantheism, in my judgment, is a form of theism, I could not hold that theism must necessarily be over-worldly in all relevant senses;

and it need not be "super-natural" unless in contrast to a "nature" that, according to theists, is really de-natured. Similarly, since the hypothesis of a limited deity does not seem to me to be necessarily a covert denial of theism, I could not claim more than that the God of theism must operate on a scale that is not sub-cosmic. He must be big enough for that or else be a mere godling. That would have to be affirmed, I think, even by those who would say with Shelley in *Prometheus Unbound*:

> How glorious art thou, Earth! And if thou be
> The shadow of some spirit lovelier still,
> Though evil stain its work, and it should be,
> Like its creation, weak yet beautiful,
> I could fall down and worship that and thee.

We have therefore to deal with a vast canvas. We may be able, indeed, by an effort of sophisticated intellectual imagination, to speak as if the physical cosmos was only a petty department of the nobler opulence of Being Itself. People say and mean these things, and they may be right. But cosmological theism is still a very big subject.

In the main, I have attempted to follow in the wake of the cosmological argument. This, according to Dr. Webb, might be described, without serious error, as the argument of ancient philosophy,[1] but it had the support of Leibniz and of other celebrated writers in less ancient times. It may be stated in various forms, but the form that seems most notable is that in which a certain existential premiss is affirmed, and theism is inferred by an immediate inference by complementary relation. The existential premiss is really complex. It is that there is a world, but that the world is contingent. If necessity be complementary to contingency, the existence of a necessary being is inferred from the admitted existence of a contingent being, the world.

I am not attempting to dispute the proposition that there

[1] *Problems in the Relation of God and Man*, p. 165.

is a world although, as I have said, doubts may legitimately be raised regarding the sense in which we may be said to know for certain that there is a single entity that we should call *the* world, *the* cosmos, *the* universe. Our difficulty was rather with the other part of the double-barrelled existential premiss asserted. Can we be certain that the world is "contingent" in any sense that implies a specific over-worldly complement? Is it, for example, an effect which implies a cause? Is it incomplete in a sense that implies an overworldly complement at least as great as itself? Is it, so to say, only a selection from the alphabet of reality whose Alpha and whose Omega is God? Such "contingency", in the relevant sense of contingency, seemed desperately difficult to establish.

We tried to be rather assiduous. Is the world self-insufficient in the way that implies a creator? Must its temporal character be completed in eternity? Must there be a divine dimension of its immensity? Is it, strictly speaking, powerless and therefore dependent on over-worldly omnificence? In asking all these questions, our true intent was to examine the sort of self-insufficiency that the world was supposed to evince. Is it really self-insufficient in any of these senses? Might it not be all that there is? Is there incontrovertible evidence that *any* existent being, or the totality of existence, or God himself, could surmount and annul any or all of these types of alleged self-insufficiency? Could God, for example, be other than temporal? If his heaven were a Utopia, could it also, in Renouvier's term, be a Uchronia?

All these arguments are empirical in a certain sense, for the double-barrelled existential premiss on which they depend is held to be plain matter-of-fact. In many respects, however, the Argument from Design, with which we ended our survey, is more industriously matter-of-fact than the others we considered before it. Dr. Webb calls it "the plain man's argument",[1] and the plain man, although often most dogmatically anti-empirical, prefers a strong empirical seasoning in all his

[1] *Loc. cit*

philosophical dishes. Indeed, if the Argument from Design be developed, in the usual way, upon clearly particulate lines, that is, if it relies upon the cumulative evidence of what it regards as the multitudinous marks of planned adaptation within the world, it is empirical in the ant-like fashion which Bacon himself condemned as a bad empiricism.[1]

The Argument from Design might indeed be oecumenical, or more than oecumenical, rather than particulate. It might rely upon the apparent design of Nature as a whole rather than on the apparent design, say of a lion's claws or of the human thumb. That is not the usual form of the Argument from Design but it is a possible form, whose logic would be a form of the cosmological argument. The entire wide world itself would be said to be a planned thing, and an over-worldly designer would be inferred by an inference declared to be merely complementary if not even tautological. Nevertheless a highly-specialized type of cause would be reached. The plain man likes the argument just because it seems so very definite. It seems to be getting him somewhere, and not to be tantalisingly abstract and general. In it he has something to grip. The argument, if it were sound, would tell him more about his God, in certain significant ways, than the others do. It has also a deeper empirical dye.

Our finding was that no conclusive demonstration of the truth of theism could be obtained on any of the lines we examined. That, I may be told, might have been anticipated. Centuries have passed since very many, very large or very salt tears have been shed over the indemonstrability of divine existence and of the divine attributes. The devout among mankind have usually preferred to walk by faith and not by sight, to possess the substance of things hoped for without any appreciable interest in the irrefragability of the logic of certain doctors of theology. Indeed surprise may be shown that anyone should seriously consider attempted demonstrations of theism in this twentieth century.

[1] *Novum Organum*, I, xcv.

Examination of Cosmological Theism

To that I reply, in the first place, that refutations may become antiquated as well as proofs. It is always dangerous to take them for granted. In the second place, these venerable arguments appear to me to explore the central part of the logic of the situation. Their failure, if they do fail, may therefore be highly instructive. In particular, a close attention to them should help us to avoid a fallacy whose menace grows in proportion to the readiness with which natural theologians are prepared to ignore the requirements of strict demonstration. That is the fallacious practice of reviving discredited demonstrations as if they were near-proofs that should incline towards assent although they do not coerce an enlightened judgment. To avoid the fallacy we have to insist that the failure of a demonstration *is* a failure. It need not indicate that the searchers are getting warm, although their quarry, for the moment, is still elusive. It is worth while being patient about the traditional demonstrative proofs of theism if, by that means, we avoid the airy vacuity of assuming that inconclusive proofs raise a presumption in favour of their professed conclusion.

On the other hand, it is clearly *possible* that where demonstrative proof is absent, reasonable conjecture may still have an important place. Problematical or hypothetical theism, therefore, might succeed where demonstrative theism failed. In cosmology the theist's venture, or some similar venture, may be credible if not indispensable, and perhaps more credible than any other. We have to pay at least equal attention to this aspect of the problem.

Here I have pleaded for a *distinguo*. I have, for instance, said nothing against pantheism as a description of the world, nothing against the philosophy of *Deus sive Natura*, of a sanctified or deiform cosmos. I can see no way of showing, and I do not in fact believe, that these high-flying descriptions are simply misdescriptions. If it be said that "Nature", in ordinary speech, or in the jargon of science or of philosophy, is denatured when it is represented as being godless, I think, personally, that the statement may very well be true, although it contains too many imponderables to deserve to make a deep impression upon most cosmological debates. In this sense, Nature way well appear to have many "supernatural" properties, for it is richer than a denatured Nature. The truth must restore any thefts that may have been perpetrated.

On the other hand, if the theistic hypothesis takes the form of asserting the dependence of the world upon a wholly overworldly being or set of beings, its difficulties seem to me to be prodigious, and its cosmological basis to be radically insecure. The point, it is true, may be piously obscured by sufficiently rapid oscillations on the theme of "immanent" and "transcendent". In this way pantheism may be blessed and may also be banned in the most eclectic fashion, but I cannot see any hope for philosophical theology when such devices are freely employed.

As we saw, one of the major difficulties that attends a wholly transcendent theism is the question of temporal succession. It is impossible that deity should be effectively related to any succession without itself containing succession. It must vary concomitantly with its diverse productions; it must last at least as long as its products. The question may indeed be raised whether there is a single measurable cosmic chronometry; but the existence of succession is one thing, its measurable chronometry quite another thing. Even the illusion of succession implies some succession somewhere. Consequently the problem cannot be evaded by the overweening assertion that succession is only phenomenal, not "absolute" or "real". If there seems to be change there is *some* change, even if the change be the change in an illusion. And such changes are successive. I do not say that the concept of existence itself implies succession, but I affirm, like everybody else, that we do not use the word "existence" accurately or scrupulously unless we believe that anything that, as we say, "exists" has, directly or indirectly, a determinate place in a process of succession.

Examination of Cosmological Theism

Accordingly a non-temporal transcendent deity, if he could exist, a deity who neither varied nor endured without variation, could neither create nor sustain the temporal universe, and would be wholly outside the ambit of all intelligible cosmology. There could be no bridge to unite him to empirical data, after the fashion, say, of the Argument from Design. No arch could span the gap between a self-insufficient world and his too ample self-sufficiency in the way that is taught by the Cosmological Argument.

The idea of a deity who transcends all space is differently circumstanced from the idea of a deity who transcends all time, and for two reasons. In the first place, space may be less stubborn than time and change. Even the illusion of change implies the existence of *a certain* change, but the illusion of space does not similarly imply the real existence of *a certain* space. Seeming space, as in a dream, does indeed show the fact of *seeming* spread-out-ness or extension, but that appearance might be illusory. In the second place, human ingenuity has busied itself for many centuries over the problem how a non-spatial mind might effectively control, or appear to control, a spatial organism. This ingenuity may have been vainly expended, and the situation is still very involved; but it would be the worst sort of dogmatism to assert that no views of the kind deserve a hearing.

Let us examine these two questions separately.

If space were illusory, cosmology, in any ordinary sense, would be illusory too. Cosmology is discourse concerning the wide world. If the world were not truly wide, what would have to be explained would be the illusion of its width, that is to say, the dream of a wide dream-world. Certainly if what we call the "world" were a *phenomenon bene fundatum*, it would not be a dream in the sense in which we contrast dream-experience with waking experience. The "phenomenon" might indeed be so very well grounded that the distinction between phenomenon and reality night not be much more than titular or official. Nevertheless, *ex hypothesi*, it would be mixed with illusion, and to that extent, cosmology, in its ordinary sense, would be illusory too.

Some philosophers hold that the existence of an illusion implies the existence of *some* reality characteristically similar to the illusion. In the common way, we do make this assumption. If anyone, for example, hears ghostly voices, we do not usually infer that there *are* ghosts, but we do infer that *voices are* sometimes real, and we should draw this inference with confidence even if the voices called ghostly were in our own ears only and were not foreign voices. I do not think, however, that a general common-sense presumption of this order can be converted into sound metaphysics by any sleight of philosophy. There is no metaphysical impossibility in the supposition that vast tracts of our *prima facie* experience are infected with an illusion that is intellectually but not imaginatively corrigible. I shall return to the question. Meanwhile I am content with the reflection that if space be illusory, so is cosmology. The argument from an illusory cosmos to its genuine deity is profoundly different from the argument from a genuine cosmos to its genuine deity.

Regarding the second question, I allow that when we refer to existence in the ordinary way, we refer to what we take to be a single executive order, spatial, temporal and causal. Any apparent matter-of-fact, such as a dream or an after-image, that does not, on the face of it, square with this executive world-order, is held to be indirectly adapted to that world-order by means of a dreaming or imaging organism that is a part of the said world-order. In short, the plain man anchors "reality" upon body, although he does not explicitly raise the question of the scientific or metaphysical analysis of "body". Regarding the mind, he is not a nullibist, although, if the question were put to him, he would probably deny that mental events were merely neural motions.

If the plain man's belief in "body" were a sound philosophical vaticination, the consequence would be that all reality, directly or indirectly, is bodily reality. Here, however, the

Examination of Cosmological Theism

qualification "directly or indirectly" leaves room for many manœuvres. There is no express assertion that souls and minds and thoughts and dreams are themselves literally bodies. All that is said is that they must be bodily in some perfectly definite sense or else be nothing at all. Such a statement is consistent with many theories of the mind-body relation, and might apply to the mind of the world, or to a mind-controlled universe, as well as to the mind of Thomas Jones or to Jones's mind-controlled body. Not all such theories imply the literal spaciousness of minds. What they do imply is that if there were no bodies there would not be any minds, and that if there were no bodily world there would not be a God.

Let us continue this discussion of the transcendence of deity in relation to cosmology.

One of the principal obstacles that beset all arguments from the world to God is the doubtful legitimacy of arguing from relations or connections *within* the cosmos to a similar relation or connection *between* the cosmos and some transcendent entity. These difficulties are peculiarly serious if the relation is supposed to be causal, and they are quite exceptionally serious if the causal relation in question is held to be some particular type of causality such as voluntary design. Indeed, if the truth be, as many philosophers hold, that causality is always indemonstrable in general and problematical in special, an alleged *demonstration* of God's transcendent causality would be forever impossible.

To speak generally, however, there is nothing illegitimate in supposing that a principle discovered *within* the world may also apply *to* the world. The question in that case is whether there is any need for such a supposition, and whether an attempted explanation along these lines should be supposed, on the whole, to have more explanatory virtue than any other that may be opposed to it. Here Humians, Kantians, old-fashioned positivists and the new logistical positivists have a short and easy method of confounding theological cosmologists. Our sole evidence of existence, they say, comes from sensation and must remain sensory. That exists which can be sensed, or at least (to allow us e.g. to infer the existence of the other side of the moon) is of the order of the sensible. There is no other meaning, they say, for an "existent"; and certain modern authors would add that since the past is irrecoverable, the only possible meaning of any hypothesis concerning existence is that it should be verifiable in some *future* sensory experience. The cosmos, however, cannot be sensed, and neither can a transcendent deity. Theological cosmology is therefore a vain thing. Even if, by excess of charity, it could be allowed some vague sort of meaning, it could neither be supported nor refuted by any appropriate test. Consequently hypothetical or problematical theism belongs to fairyland. That is what these people say.

I want to make it clear that I am not using this argument. I do not in fact accept it. I have maintained, it is true, that space and time, so essential to all cosmology, do have distinctive properties that are revealed in sense and cannot be wholly logicized. I have also maintained that the cosmological and the other arguments with which this first series of lectures is concerned, rest, in part, on an empirical premiss, and therefore differ in principle from the Ontological Proof to which (among others) we shall proceed in the next series. I also allow that, in calling this existential premiss "empirical," I am relying, in the main, upon the basis of sense-experience. I cannot, however, accept the view that "to exist" and "to be sensible" necessarily mean the same thing. I would not even say that without sensation there could be no evidence of existence. All I am saying is that our belief in the existence *of the world* does involve such evidence and involves it very extensively. Again, I cannot accept the dogma that futurity is the only possible home for all legitimate existential enquiries because it is the future, and it alone, that verifies. This opinion seems to me to be plainly false. Such verification, when it occurs, must be *present* (or just past), not future; and all our empirical evidence is at least as much remembered as present. Moreover the past is not always "irrecoverable" in thought. It is recovered, I hold, in memory. What is inevitably irrecoverable is a *present* past, that is to say, a piece of nonsense.

Accordingly, we need not do obeisance to this kind of radical empiricism in our philosophy, when, as cosmologists, we argue from the empirical premiss that the world exists. Again, as I have recently been contending, we may not say that the attributes we ascribe to the world exhaust the attributes of reality. Reality may have an infinity of other attributes than those we are accustomed to ascribe to body. What is more, the argument from the world's existence does not prohibit the view that there may be an over-world, or an over-worldly divinity, or many of these. Most pantheists seem to assume that the world, whatever else it may be, must at least be the body of all reality. That is the interpretation that anyone would naturally give to the philosophy of *Deus sive Natura*, and I am not asserting that the theory is false. I doubt, however, whether it could be proved, and I do not think it should be assumed to be true. I have argued, indeed, that the attribute of time (when carefully distinguished from measurable chronometry) does permeate all conceivable reality, and I do not doubt that all reality is also spatial. Therefore I incline towards the view that Nature is God's body, if there be a God. I do not suggest, however, that the point could be convincingly established beyond all reasonable cavil, and have therefore admitted that there might be an over-worldly supreme being, who could not be timeless but who need not be spatial and therefore need not have any body at all.

Let us return to the double-barrelled existential premiss of the Cosmological Argument. The existence of the world is the fulcrum of cosmological theism. We may say, if we choose, that, except for the time in it, the world might not have existed. But it does exist, and supplies a part of our duplex existential premiss. We may call its existence "brute fact" or may try to show that the fact is not so very brutish after all.

Brutish or not, however, it is fact. It need not be the whole of fact. It may be relatively trivial. It may be unsubtly pedestrian; but we cannot conjure its existence away. To speak of it slightingly as "brute" fact, sense-evidenced fact, "opaque" fact may sound very elevating, but need not be more than empty speech. Philosophers may have to disintoxicate themselves from the ether of such verbiage.

What are we to say, then, about the second arm of the double-barrelled existential proposition on which the Cosmological Argument rests? i.e. about the assertion that the existent world is self-insufficient in a way that implies an existent complement.

For the most part the required complement is described as a cause—a First Mover who moves "as desire moves"[1] according to Aristotle's honey-pot argument, a designer who moves as a mechanic moves, a musician who moves as a conductor moves, or the like. In other words, cosmology is supposed to imply cosmogony. Nature is generable and must have been generated in some non-natural way.

Here it is necessary to distinguish between the scientific and the metaphysical aspects of the question. Let us begin with the former, and consider it rather fully since it has, so far, not been discussed at all in these lectures.

We speak of the evolution of the solar system, of the galactic system, and of the astronomical universe itself. That is scientific cosmogony, and although it is largely guess-work, it is quite a different sort of guess from old-world stories about Uranus and Gaea, or the spider that some Easterns credited with the spinning of the world. The likely tale that astronomers tell us is to the following effect. Kant guessed in an early work[2] that the galaxy and the nebulae had developed from an almost uniform diffusion of matter throughout space, and he suggested a theory of gradual astronomical development in place of the sudden creation of the stars. His physics,

[1] *Met.* Λ, 1072 b 3.
[2] *General Natural History and Theory of the Heavens*, 1755.

however, was defective in so far as he postulated a spin that a self-contained system could not generate. Laplace,[1] a more expert physicist, put forward the more plausible theory that the solar system had developed from a hot nebular mass in rotation, which shrank as it cooled but rotated the faster. Rings of matter were thrown off which, cooling, became our planets. The main technical objection to this hypothesis is that it implies that twin-stars should invariably have been formed (as in some other solar systems). Since that did not happen in our own solar system, the more probable theory, in its case, would be that a "rogue" star tore a great filament out of the sun.

That is, or was recently, the fashionable theory of the origin of our solar system. A nebular hypothesis regarding the evolution of the present astronomical system is sustainable. A gaseous distribution with a sufficient spin would tend to aggregation in denser bodies, and the celestial clotting would go on.

Such theories involve immense extrapolation, but they do give a picture of the way in which the present astronomical cosmos may have come to be what it is, having once been very different indeed from what it is now. Such pictures form a series which suggests that the book of starry Nature has its title-page and its colophon. In its beginning, the present astronomical system was almost uniformly diffuse. An end, however, is also foreshadowed. The system tends towards a condition in which no energy is available for work. Entropy, beginning at a minimum, tends towards a maximum. This tendency may be retarded, e.g. by radio-active forces; but it is not completely stayed.

> Tristis item vetulae vitis sator atque vietae
> Temporis incusat momen caelumque fatigat
> Nec tenet omnia paulatim tabescere et ire
> Ad capulum spatio aetatis defessa vetusto.[2]

[1] *Système du Monde*, 1796.
[2] Lucretius, *De Rerum Natura*, end of Book II.

Theism and Cosmology

More generally it is argued that the running-down of the astronomical universe towards a condition of heat-death in which no work can be done is as good as certain despite desperate suggestions of succour from cosmic rays, or from the regeneration of hydrogen atoms, say, in a thunderstorm, and that anything that will have an end must have had a beginning. Philosophers say this as well as physicists. "If I believed in the second law of thermodynamics as much as Sir Arthur Eddington does", Mr. Braithwaite has said, "I should feel impelled to believe more than he does in an 'incredible' First Cause."[1] Brentano, that very acute philosopher, was quite positive about the principle.[2]

It seems to me to be arbitrary. Let it be granted that the physical universe is moving towards a condition of maximum entropy or of minimum organization. The alleged conclusion is that it must have been moving from a condition of minimum entropy or of maximum organization. This does not seem to me to follow. My body is moving towards death, i.e. a condition of minimum organization, and may be said to have been slowly dying for many years. It does not follow that it was most alive when it began. I allow that in fact my body had a beginning, but that would not be proved by the simple circumstance that it will have an end. More generally, and using ordinary theological language, we may say that if an omnipotent deity created the world, he might decide never to annihilate it, *and* also that if the world, *a parte ante*, had been co-eternal with its creator, he might nevertheless decide and would be able to annihilate it. There is no contradiction.

In any case, maximum entropy is not annihilation. By parity of reasoning, therefore, a condition of minimum entropy need not be an absolute beginning *ex nihilo*. I do not even see that such a beginning need be abrupt, although that is the conclusion that Sir Arthur Eddington dislikes but does not see how to avoid. "Philosophically", he says, "the notion of an

[1] Quoted by L. S. Stebbing, *Philosophy and the Physicists*, p. 258.
[2] *Vom Dasein Gottes*, p. 395.

Examination of Cosmological Theism

abrupt beginning of the present order of Nature is repugnant to me, as I think it must be to most; and even those who would welcome a proof of the intervention of a Creator will probably consider that a single winding up at some remote epoch is not really the kind of relation between God and his world that brings satisfaction to the mind. But I can see no escape from our dilemma."[1]

Indeed I submit that these arguments may be embarrassing to those who believe that what we have discovered about the present pattern of the astronomical universe is an adequate description of the sempiternal pattern of all physical reality. Moreover, they do not establish anything approaching an absolute beginning or an absolute end of bodily existence, and they do not show that the present pattern of the astronomical cosmos was without physical antecedents, and required the intervention of non-physical or of hyper-physical causes.

In this matter, I do not find the attitude of such of our modern theologians as are competent to form an opinion very helpful. Take, for example, the views recently expressed by that most competent person, the Bishop of Birmingham. "The God of the trigger", he says, [i.e. the God who intervenes at critical moments] gives no appreciable satisfaction to his reason.[2] His episcopal eye is cold when he thinks of this spinner of a flaming cosmic top, and he consoles himself with the doubt whether "inference will ever allow man to know the physical circumstances of the beginning of the cosmos with the certainty with which, for example, he can infer the positions of the planets five thousand years ago".[3] That same episcopal eye, however, brightened visibly when it contemplated the modern theory that the solar system began when a rogue star tore out the sun's entrails. That, the Bishop said, might look like chance, but the probabilities were "that other similar encounters have produced a vast distribution of life-

[1] *New Pathways of Science*, p. 59. Quoted Stebbing, *loc. cit.*
[2] *Scientific Theory and Religion*, p. 409. On p. 240, however, the God of the trigger seems to be welcomed. [3] *Ibid.*, p. 408.

bearing planets throughout space."[1] The supposed rogue was therefore a decoy, "the handmaid of design". It was a gracious vagabond.

The earlier among these statements seem to me to be essentially sound, the later ones to be a flight of devout fancy.

Indeed it seems to me to be clear that scientific cosmogonies regarding the present condition of the physical universe can have little bearing upon natural theology. That may be a somewhat belated defence of the omission of such arguments in the earlier lectures of this course; but, such as it is, I offer it now. I am not suggesting that science is irrelevant to cosmological theology. On the contrary, I believe that all philosophical theologians should adapt to the instance of science what Sir John Eliot said about parliaments. "None ever went about to break science, but in the end science broke him." Here, however, there is no question of setting about to break science.

If the evolution of the astronomical cosmos from a uniform distribution of hot, spinning, nebular gas were a proved fact like the punctual reappearance of Halley's comet in the last few millennia, and if the gradual development of that cosmos towards a condition of maximum entropy were similarly proved, cosmologists would have to accept the situation, whether their leanings were theological or anti-theological. As matters stand this theory appears to be the best conjecture that there is, making plausible extrapolations with a good deal of empirical backing. Therefore it is a good deal more than a flight of scientific fancy, but it is very far indeed from being an invincible datum for science, philosophy, theology or anything else.

Accordingly we need not examine very closely whether scientific cosmogonies of this kind would or would not accord with certain types of theism. Such a cosmogony, I think, might be rather more embarrassing to certain types of atheists than to most types of theists; but not very much. And neither

[1] *Scientific Theory and Religion*, p. 404.

Examination of Cosmological Theism

party need be reduced to despair. Theists may welcome the "God of the trigger", who has to load, and to re-load, the gun. The gun, they say, is not the All. It requires a greater being behind the gun; and that greater being might be immaterial. (If he is omnipotent he could certainly load the gun.) On the other hand, the argument does not establish the need for an immaterial God, and it infers the existence of an overworldly being only upon the supposition that the "world" means the present pattern of the physical cosmos. Materialists might attempt to meet the argument by reviving theories of cataclysmic as opposed to gradual evolution, and indeterminists would be quite happy about the whole business. In modern times, when indeterminism and the possible annihilation of matter are freely suggested, it is hazardous to set bounds to the speculative vistas that such ideas encourage. If, however, the argument were held to show that the world could not be its own trigger, it might still be doubted whether the hand on the trigger was likely to be divine. By hypothesis the God on the trigger allows heat-death to occur before he re-loads the gun. That is not a cheerful plan, and so far as the argument goes, there is no suggestion that the next discharge will lead to anything except a repetition of the first. There might just be cycles of gloom, none of them lasting for ever. The theists, therefore, might well find the argument embarrassing.

But let us return to the wider philosophical questions that are our proper business and, in particular, to the supposed First Cause or Source of the world. What are the philosophical reasons for holding that cosmology implies cosmogony, that the world cannot be ultimate and the like?

According to theists, *something* is ultimate, namely God, and can be discovered to be so. This part of theistic theory was jealously retained, in the theistic tradition, even by authors who, like John Locke, were rather slipshod in their proofs of theism. Locke denied that *everything* must have a cause. That proposition, he said, was "not a true principle of

Theism and Cosmology

reason but the contrary". What was true, and indeed, certain, was "that every thing that has a *beginning* must have a cause". God himself, having no beginning, had no cause.

Similarly it was denied that every cause was also an effect; for the First Cause was not an effect. It was affirmed that the Great Designer designed but was himself undesigned; and so forth. According to theism, therefore, it is impossible to go beyond or to go behind God, but it is necessary to go behind or to go beyond the universe, unless, as pantheists aver, the universe, in some sense, *is* God. An agnostic holds that there is no way of determining, either with certainty or with the tiniest atom of probability, whether there is anything behind the world, and no possibility of intelligible debate concerning what such a being could be. An atheist denies that *God* is behind or beyond the world. He probably believes in the world's self-sufficiency, but he need not commit himself upon that question. Theists, however, speak with finality about the world's self-insufficiency and would not be theists if they wavered.

Consequently the self-insufficiency of the world is an essential step towards most forms of cosmological theism. If the world were self-sufficient, there would be no ground for denying its ultimacy. The second barrel of the double-barrelled existential premiss upon which the cosmological argument rests is therefore of the greatest moment. If the world's self-insufficiency is taken to mean, or to include, its causal dependence, theistic cosmology implies the truth of some sort of cosmogony, some origination of the world. That is the commonest argument of the kind; but the world's self-insufficiency might be interpreted in other ways, and these have also to be examined. If all of them are denied, the case for cosmological theism, except as a form of pantheism, would seem to be abandoned.

I tried, in the third lecture of this series, to examine generally some of the principal senses in which the ultimacy of the

Examination of Cosmological Theism

world was denied. I must now be briefer, and, of course, I could not claim then (and I cannot claim now) that I can give an exhaustive survey. I think, however, that the chief of these reasons are the following: The world is said to be self-insufficient, either because it is illusory, or because it is fragmentary, or because it is other-determined. These contentions need not be independent. As we all know, certain philosophers hold that whatever is fragmentary is, as such, partially erroneous, and therefore in some sort illusory. On the other hand, these several arguments have a certain independence, and I propose to treat them separately here, putting aside the question of their possible intermixture.

The reasons adduced for holding that the world is illusory are mostly indirect. A favourite argument is that if the world were not illusory it would be an actual or completed infinite at any moment; and that, it is said, would imply a contradiction. Another common argument is that sensation is incurably superficial; that the sense-evidenced, consequently, cannot be ultimate; but that the world is sense-evidenced if it is evidenced at all.

(1) To the first of these arguments I would reply that an actual infinite does not imply a contradiction, and that, if it did, theism would not be more fortunately placed than the contrary of theism. If an actual infinite be a contradiction, it is not less of a contradiction when the infinite being is said to be divine. Again, as I have argued repeatedly, the supposed difficulty plainly applies to time (which grows and so is never, in all senses, infinite). If deity be effectively related to a changing and temporal cosmos, he cannot himself be unchanging and timeless.

Time and change, in short, cannot ultimately be illusory. Therefore the illusoriness of the world cannot be proved by any argument that is based upon the logical implications of time and of change.

On the general question there is a pleasant story in Buddhistic theology:

Theism and Cosmology

There was a monk indulging, against the teaching of the Master, in cosmological enquiries. In order to know where the world ends, he began journeying far away in the sky, interrogating in succession the gods of the successive heavens. The gods "Servants of the four kings of the cardinal regions" said to him "Ask the Four Kings"; the Four Kings said to him "Ask the Thirty-Three Gods". . . . Finally the Great Brahma himself became manifest, and the monk asked him where the world ends. Brahma answered "I am the Great Brahma . . . the father of all that are and are to be." "I do not ask you, friend", said the monk, "whether you are indeed all that you now say. But I ask you where the four great elements—earth, water, fire and wind—cease leaving no trace behind." Then the Great Brahma took that monk by the arm, led him aside and said: "These Gods, my servants, hold me to be such that there is nothing I cannot see, understand, realize. Therefore I gave no answer in their presence. But I do not know where the world ends. . . ."[1]

(2) The argument from the sense-evidenced character of worldly existence seems to me to be quite decrepit. I allow that spatiality, temporality and many other worldly properties are revealed to our senses in a rudimentary way, and, apart from our senses, may not be revealed to our intellects in any way at all. They would escape a pure intellect, human, angelic or divine, just as much as colour would, or the taste of a pine-apple. This holds of succession, which cannot be logicized without remainder. But succession is metaphysically invincible. Therefore the sense-evidenced need not be illusory. Therefore the world need not be illusory on the mere ground that it is sense-evidenced in certain of its essential features.

Similar arguments apply to the contention that the world is fragmentary and therefore self-insufficient. At any given moment in the past it *was* in a sense fragmentary, for there was more of it to come. The same applies to God's existence at any past moment. He also has been fragmentary in that

---

[1] *Dialogues of the Buddha.* Quoted by Poussin, *The Way to Nirvana*, p. 105.

Examination of Cosmological Theism

way. Of the future we cannot speak with precisely the same metaphysical confidence. If matter, conceivably, might be annihilated, so might anything else, including spirit, human or divine. Every past moment of existence, however, has been a moment in parturient process. There is no reason for denying that such parturience will continue without end. Any contrary supposition is as "repugnant" to most of us as it is to Sir Arthur Eddington. The fragmentariness of worldly existence, therefore, is not a proof that the world is illusory, unreal, or, as some philosophers would say in an indefensible mode of speech, "real" in some subordinate sense that is not "ultimate".

I should further remark that if the world be not illusory and if it is a fragment, then its complement, even if divine, must also be a fragment, perhaps bigger than the world and perhaps smaller; but still a fragment. The complement would not be the whole. It would be a complementary fragment. Therefore the argument from fragmentariness would destroy God's self-sufficiency (if the argument were true) as well as the world's self-sufficiency. It would be flatly inconsistent with any form of theism except totalitarian pantheism.

(3) We have already discussed the supposed causal dependence of the world in many aspects of that question. The aspect that affects us now and is specially appropriate to a First Cause in the narrower interpretation of that term is the common idea that the world could not be an ultimate piece of existence because it is causally dependent. That argument depends upon an equivocation. If derivative be defined as the contrary of "ultimate" then an effect is never ultimate. On the other hand, an effect has not a derivative *mode of being* when compared with its cause. Both are pieces of existence in the same final sense. Neither, in any intelligible sense, *exists more* than the other. A similar comment should be made upon the idea that "ultimacy" involves some kind of necessity, and that the world, for all that we can see to the contrary, *might not* have existed. If the world were the effect of free creative volition, it would be true to say that the world might not have

existed. From the mere premiss "X exists but might not have existed", however, it cannot be inferred, by any known logical principle, that anything other than X exists.

The last point deserves very close attention. Most assertions of the world's incorrigible self-insufficiency are assertions of its imperfection when assessed according to some particular standard of "perfection". Thus it is said to be an imperfection to have parts or passions. Therefore the worldly or the human must be "imperfect" however enormously they may be magnified. It is said to be an "imperfection" to be just "brute" fact, stubborn and, metaphorically speaking, "opaque" (as any compages of sense-impressions is said to be by lucidly lucent intellectualists). According to Bradley and to some other philosophers, there would be "imperfection" if cosmology were intellectually translucent but nothing more; for the soul would be starved by any philosophy that a man could not "prove upon his pulses". It would be an imperfection, some say, if there were any plurality, any distinctions, any limitations, any negations, any finitude in reality. The "ultimately" real must be a pure simplicity or unicity.

Men may certainly form such standards, and they may speak (if they choose) of an ultimate or superlative "reality" that conforms to the standards. Similarly (if they choose) they may define this superlative "reality" in such a way that all that is manifold, partitioned, untidy, compounded or "brutally" factual is excluded from it. What they cannot do with impunity is to compel other people to admit that nothing can be real except such a superlative "reality". Anyone who can affirm, without illusion, that there is an X, is affirming X's reality, and there are no degrees of reality. Either there is an X or there isn't. To prove that there is something X-like that isn't X is not to prove that there is an enfeebled or "unreal" X.

Therefore to say that the world is "imperfect" in various senses is in no way to deny its reality; and to admit its reality is to admit its "ultimate" reality. If, in any instance, X exists, that of itself is ultimate, and, because it is ultimate, there is

Examination of Cosmological Theism

no going behind it. Therefore the primary question in all cosmological arguments is whether there is or is not a world. There is no difference between saying that there is a world, and saying that there "really" is one. It is also true, no doubt, that if we are entitled to say that there is a world, we must be able to say so without illusion; and I am not maintaining that there cannot be pervasive and persistent illusions about what we call the world. In order to be entitled to do so I should have to refute scepticism in nearly all its forms, and I have not embarked upon that enterprise. My view upon the matter is that while nearly any existential assertion may be corrigible in detail, the whole body of such beliefs cannot seriously be regarded as moonshine. That in itself might not suffice to rescue the world from a determined doubter. I should maintain, however, that the body of our beliefs about the world while alterable in most small matters and (very likely) in some pretty big ones, is immune from a general intransigent nihilism, and also from a nihilism that is just a little milder. If we cannot believe that there is any sense in which there is a world, our capacity for belief is too attenuated to make discussion profitable.

When I say that it is illegitimate to attempt to go behind any piece of existence, I mean, quite simply, that existence is ultimate. I do not mean that one piece of existence may not be behind another piece of existence. Even in the literal sense, Polonius was behind the arras. It was he that made the arras twitch, and if it had not been for such as he (that is to say, statesmen and courtiers) there might not have been any tapestries at all. More generally, much that exists may be relatively trivial and flimsy, concealing what is weightier, more powerful, better, more durable than itself. To admit that the world exists, therefore, is not necessarily to deny that it is weak and trivial and ephemeral. It may be asking rather little of God when we insist that he must at least be capable of operating upon a cosmic scale. (But, as I have said, I think it is asking quite a lot.)

Theism and Cosmology

Our present theme, however, is the relation between cosmology and theism. The world is our oyster, a dumb witness. We are asking whether the witness is a witness to deity. Therefore we attach some importance to cosmological testimony, particularly since the witness, being dumb, cannot wilfully lie. Cosmological evidence may not be all the empirical evidence that we have—for mystics and some others rely upon a species of personal communion with deity that may have little or no cosmological import. Yet there is no spiritual experience (so far as we know) that does not have its place in cosmological temporal process and is not connected, at least indirectly, with a physical body that plays its part in the physical cosmos.

Therefore we should be wary. We do not want a flat brew and we do not want a heady brew. I shall try to discriminate between what I call flatness and what I call headiness in this connection.

I think the brew would be flat if we argued as follows: By interrogating our experience, chiefly sensory, we arrive at the galactic universe, and have reason to believe that there are other galaxies beyond the Milky Way. This we call the physical universe or the "world": and it exists. It is "Nature" and "Nature" exists. Therefore we may assume that "all that is" is at least natural, and that all our enquiries must be conducted within this natural framework. Therefore, if there be deity, he also is contained in this natural framework. All existence is "natural" existence. There is no "other" world. There is no supernatural transcendent realm, no *Zweites Reich*. The world of nature may indeed have over-worldly characteristics, omitted in the secular descriptions of it. Neglect of its deiformity may lead to an inadequate description of it; but there is no non-natural existence, and all natural existence has its astronomical framework.

I call this brew flat, but not because its conclusion is false or its standards quite unimaginative. On the contrary, it seems to me that theism has more hopeful prospects along these

Examination of Cosmological Theism

lines than along most others, and I do not find any lack of imagination in this philosophy of *Deus sive Natura*. The flatness, in my opinion, is due to the *assumption* that the thing must be so, that because "nature" exists, therefore an "other" world or a transcendent God must be mere empty speech. That, I submit, does not follow. It is true that if God be effectively related to the physical cosmos, he and that cosmos must form some sort of unity together. It does not follow that God is only an attribute of "Nature". I have argued, indeed, that there must be co-variation between God and the world (however transcendent God may be) and so that a certain temporal connection must be affirmed; but the degree of community that can be proved in this way is not very extensive.

What I call a heady brew, on the other hand, is the swift transition to a transcendent theism from the alleged precariousness of the world itself. Unless the world is radically illusory there is no such valid transition. The world may be sense-evidenced, an Other that the intellect always finds dark, as much a jungle as a garden, and so forth. If, however, it exists, *some* existence must be sense-evidenced, intellectually darkened, wild and irregular as forest or moor is wild and irregular. It will not be less so because it depends on God. On the contrary, unless God be a limited God, these "wild" and "brutal" facts are less comprehensible with the theistic hypothesis than without it. If we renounce theism, these so-called "imperfections" are ultimate fact that has to be accepted like the fact that we were born or that our bones grow brittle with age. There is then no logical difficulty and there is no metaphysical impossibility. If we accept theism, and also accept the world's reality, we cannot deny these same brute facts, but have to combine them with the assertion that they are the products of a perfect and supernally powerful being who had produced an "imperfect" world when he need not have produced anything imperfect or anything at all. The difficulties are immense.

Therefore, although we should not *assume* the falsity of

Theism and Cosmology

such a view, it seems to me that a theistic cosmologist is likely to be much more persuasive in his natural theology if he holds that the world in some sense is divine, (that being all that he means by his theism) than if he tries to argue to a transcendent theism on cosmological grounds. He has then, it is true, to admit that if space and time and other ineradicable features of the world are marks of imperfection, God too, in that sense, is "imperfect". The traditional attributes of deity in Christian theology must therefore undergo a certain revision, and the same would be true of the other great religious theologies. It is true, again, that if our cosmologist holds with the late Archbishop Söderblom that "the taboo, sacred, holy is the fundamental idea of religion"[1] he could not very well be a pantheist. For the sacred, in this sense, is something set apart and the totality of things cannot be set apart. In a wider sense of holiness, however, he might say that all is holy, and, like another archbishop and Gifford Lecturer[2] take a "sacramental" view of Nature. What he has, then, to hold is that our current descriptions of Nature are, not false, but inadequate. They leave much of its divinity out; and the features thus omitted may be restored by a sound philosophy or by a reasonable theology. Cosmologists need have no objection in principle to such a piece of metaphysical irredentism.

It is, of course, quite a different question whether the deiformity of nature itself (as opposed to the signature of a supernatural deity) is a more accurate description of nature than any other. In that connection we examined the argument that Henry More used against Descartes, namely that if "nature" were merely mechanical it could not have the cohesion and the organization that it has in fact, unless our natural knowledge is much more sadly mistaken than we have any good reason for believing. More's was an early expression of a type of argument that is still in the fashion, and its date is

---

[1] *The Living God*, p. 20.
[2] W. Temple, *Nature, Man and God*, Lecture xix.

significant since it was one of the first effective protests against the new physics that had conquered the brave new intellectual world. Among contemporary writers, Professor Stout, in his *Mind and Matter*, is (as we saw) a prominent representative of a somewhat similar contention. On the negative side, however, arguments of this type are very far from being a sufficient support for theism. The alternatives "either mere mechanics or God" are not exclusive. There are plenty of atheistic non-mechanical possibilities. Did not Aristotle say that the world was superhuman and marvellous but *not* Divine?

Henry More himself attempted to unite two arguments, namely the argument that the cohesion and organization of the universe is the ubiquity of the Godhead in it, and the plain man's argument from design. While I would not myself agree with the late Professor Bowman that the supposed antithesis between mechanism and organization is "one of the strange vagaries of human thought",[1] I am certainly convinced that theology occupies very weak ground if it insists, in a figure that estranges it from all sound reasoning, that adaptation or co-adaptation must either be idead in themselves or be the result of the ideas of some outside schemer. As it seems to me, theistic cosmologists should concentrate their attention, not on the indefensible prepossession that the deiformity in the cosmic pattern must have been *imposed*, must have been *put* into whatever has it, but upon the godlike character of that pattern itself. But I have said enough about that.

Nevertheless, there is much to suggest the deiformity of nature, unless these suggestions are strangled at birth by prepossessions concerning what deity must be. It may be good policy for scientific cosmologists to ignore these suggestions. They have quite enough to do if they retain a theological neutrality and they do not always better their own physique by diving abruptly into the deep waters. The same neutral policy may also be profitable in much philosophical discussion.

[1] *Studies in the Philosophy of Religion*, II, 365.

Theism and Cosmology

Philosophy, however, has room and has need for those cosmologists who brood upon the sublimity of the cosmos and its affinity with high spiritual aspiration. I am not suggesting that those who explore these great matters receive a plain answer to their questions, or that none of their questions is misconceived. I am saying only that these theistic cosmologists have a definite case, and are serving humanity (perhaps more than humanity) in exploring the implications of natural deiformity.

One remembers Kepler's prayer: "I thank Thee, my Creator and Lord, that Thou hast given me these joys in Thy Creation, this ecstasy over the works of Thy hands. I have made known the glory of thy works to men so far as my finite spirit was able to comprehend Thine infinity. If I have said anything wholly unworthy of Thee, or have aspired after my own glory, graciously forgive me. Amen."

In any case we have to note that even if the double-barrelled existential premiss of the cosmological argument could be established, namely that there is a world and that it is self-insufficient, there would be immense difficulty in further establishing that the type of self-insufficiency that the world showed entailed just one complement, and that this complement was big enough and was fine enough to be God.

There would be a very long journey from the world to a complement that might thus be inferred, i.e. to the God of Christian theology; and it is the further stages of this journey that have the greatest interest for those who belong to the Western tradition. What they ask themselves in the main, is whether this First Cause, or this Necessary Being, is holy, righteous and wise, whether there is grandeur, compassion and fineness in its being. To pursue such enquiries, however, is to ask whether these characteristics belong to the rhythm that is in the pulse of all existence. Here it need not be assumed that the deiformity of existence according to a pantheistic philosophy is inferior to the transcendent hypotheses of other theists.

Examination of Cosmological Theism

Similarly in the Argument from Design, it is not the mere fact of design (supposing that it could be proved) that kindles men's imaginations, but the character of the designer. If the fact of design were all, the devil might be the designer. The argument is welcomed because it may seem to show that God (or a divine community) is a conscious entity of great intelligence, and of hyper-cosmic power. If wickedness be folly, it cannot be supposed that such a being is diabolical—for that would affront his intelligence; and although the divinity of the Great Designer may be suspect (for design might seem to be superfluous unless there were obstacles to be overcome and something lacking that the schemer designs to fill) a certain optimism is apt to dispel such metaphysical scruples. Design is the planning of means; but it is the end that counts. Those who accept the argument refuse to believe that such supernal skill regarding the means could be conjoined with supernal blindness or folly regarding the ends. Consequently they are prepared to presume a wisdom in the designer that may be but dimly and darkly apparent in the book of nature. To that extent the hypothesis itself may encourage them to outrun the evidence. Most other theistic hypotheses offer less encouragement. Substantially, however the conclusion of these writers is that the world is too good to be godless; and there they are at one with most of the others.

The deiformity of the world, then, seems to be the most natural conclusion for a realistic limited cosmology to aim at; but I have not been arguing, in any of these lectures, that cosmological theism *must* be a form of pantheism, that there could be *no* intelligible cosmological arguments towards a supra-mundane deity. I am, however, of opinion, and would like to say explicitly, that if a cosmological theism is based not upon the illusoriness but upon the reality of the world in some defensible sense of the term "world", its contention (namely, that the existence of a godless world is an affront to metaphysics; that is, to any serious enquiry into what is ultimate) is more plausible and I think better grounded if it is, broadly speaking, of the pantheistic kind. Its natural and, I should say, its most logical form is the doctrine that the world is intrinsically deiform, even if God be much more than worldly; or, in the negative, that a secular or godless world is a world misconstrued, attenuated, exsanguine.

I should like, then, to make certain observations regarding the form of theism that is called pantheism.

Critics of pantheism, or of any form of theomorphism, commonly object to what Dr. Webb calls "the easy way of finding God in things just as they stand".[1] They may add that their quarrel is with the statement "All is God", and not with the statement "God is all". In saying this Dr. Webb claims support from Hegel and also from the text in 1 Cor. xv. 27–28: "For He hath put all things under his feet . . . that God may be all in all".

Here the objector seems to me to beg the question. Anyone who is warned against the "easy" course of accepting things just as they stand may retort, with equal ease, by asking the American question, "Just how do they stand?" There is no need for pantheism to be superficial, no occasion for it to be intimidated by the apparent Godlessness of hairs and lumps of coal, no reason for assuming that every arbitrary division we choose to make in nature must reveal the pattern of deity with equal clearness. Pantheism may be a challenge and not an evasion.

Again the statement that "God is All" *is* a statement of pantheism, although St. Paul in the context may not have meant to say so, since he was asserting God's indefeasible sovereignty over all things, even his own Son. If, however, the meaning of the statement be that God is the whole although the parts are not God, it is important to remark that pantheism need not be simply convertible with a totalitarian metaphysics, and should not be considered "easy" or rather frivolous if it renounces that particular philosophy. It is

[1] *Problems in the Relation between God and Man*, p. 257.

Examination of Cosmological Theism

consistent with such a metaphysics but is not irrevocably affianced to it.

Moreover, although a limited pantheism is a contradiction in terms, the doctrine of the pervasive theomorphism of the world might consistently be held by people who are anxious to deny that God is *equally* manifest and operative in all things.

"Every pantheism", said von Hügel, "persistently denies that the inner life of God is something far fuller and richer than the whole of His creative and providential activity."[1] That seems to me to be, very likely, a just and an important comment, but its express purpose is to take us beyond cosmology.

[1] *The Reality of God*, p. 66.

Index of Proper Names

Aetius, 67
Alexander, S., 71 ff., 182, 198, 276 f.
Aristotle, 57, 139 ff., 146, 166, 234, 257 f., 289, 302, 317
As You Like It, 150
Augustine, 28, 76, 170, 194 n.

Bacon, 36 n., 40, 260, 279, 294
Barth, 36
Bergson, 129, 164, 231
Berkeley, 180 f., 199, 215, 218 f.
Bertocci, 144
Birmingham, Bishop of, 305 f.
Blake, 39
Bosanquet, 50 f.
Bowman, 118, 317
Bradley, F. H., 50 f., 312
Braithwaite, 304
Brentano, 304
Broad, 152 n., 165
Browne, Sir T., 278
Browning, 73
Brunner, 37 n.
Butler, Bishop J., 233, 246

Carritt, 278 n.
Carroll, Lewis, 97
Cato, 66
Cicero, 28
Clarke, 45, 97
Clay, E. R., 154
Clifford, 236
Coleridge, 58
Collingwood, 123 n.
Comte, 32, 61
Cordemoy, 219
Cornford, 140
Cuvier, 103

Darwin, 129
Dawes Hicks, 277, 279
De La Forge, 219
De Noüy, 150
Descartes, 77, 134 f., 158, 176, 187, 192 f., 196, 205, 233, 272 f., 316
D'Holbach, 78
Dirac, 148, 267
Dryden, 29, 168
Dunne, 172

Eckhart, 159
Eddington, 168, 304 f.
El Greco, 124
Eliot, Sir J., 306
Eliot, T. S., 58

Farnell, 113
Franklin, 246
Freudians, 245, 258

Galileo, 289
Gassendi, 175
Gifford, Lord, 43 f.
Gotschalk, 132
Gregory of Nazianzus, 121
Grey of Fallodon, 289
Grote, J., 256

Haldane, J. S., 201
Hartley, D., 28
Hegel, 34, 43, 49, 65, 67, 278, 320
Heracleitus, 129, 163 f.
Hippocrates, 65 f.
Hobbes, 33 f., 60 f., 122 f., 159 f., 188, 192 f.
Hogben, 287
Housman, L., 263 n.

Hume, 32, 78 ff., 122, 198, 212 ff., 224 f., 231, 233, 299
Huyghens, 291

Inge, 158 n.
Irenaeus, 281
Isaiah, 116

James, W., 154, 241
Jaspers, 51
Jeans, 189 f.
John of the Cross, 159

Kant, 32, 65, 73, 76, 85, 90, 133, 162, 164, 233, 242, 257, 277, 279, 291, 299, 302
Keats, 285
Kepler, 318
Kierkegaard, 38

Laplace, 303
Leibniz, 28, 39, 46 f., 65, 77, 90, 205 ff., 283 f., 292
Lewis, G. N., 252
Liber de Causis, 209
Locke, 88 f., 175 f., 214 ff., 275, 307 f.
Lotze, 180
Lucretius, 29, 129, 303

Mackintosh, H. R., 267
McTaggart, 51, 65
Malebranche, 43, 190 f., 205 n., 218 f., 275 f.
Maritain, 74, 103
Marx, 32
Maugham, Somerset, 180
Mendel, 289
Mersenne, 76 f., 233
Mill, J. S., 38, 48
Milton, 43, 104, 123, 191, 194, 249, 274
Montaigne, 28, 30 f.
More, H., 192 ff., 261 f., 269, 316 f.
Morris, William, 263

Neurath, 32
Newton, 77, 147 f., 188, 191, 289

Nicholson, R. A., 72, 144
Nock, A. D., 33 n., 82 n.

Paley, 233, 256, 261, 270
Pascal, 77
Paton, 162
Paul, St., 36, 135, 167, 320
Perry, R. B., 240, 246, 248, 251
Pherecydes, 33
Plato, 57, 104, 183, 191, 208, 288; and see names of certain dialogues
Politicus, 77
Priestley, 175
Proclus, 85

Rabelais, 42
Reid, T., 202
Renan, 38
Renouvier, 293
Roberts, Michael, 29 f.

Sallustius, 82
Santayana, 207
Schelling, 71, 277
Schopenhauer, 234
Seneca, 66
Shakespeare, 124
Shelley, 204, 292
Smith, Adam, 78
Smuts, General, 201
Socrates, 57
Söderblom, 316
Spinoza, 67, 94, 166, 212, 233, 236 ff., 276
Stocks, J. L., 141 n., 166 n., 258 n.
Stout, 201, 241, 317

Tauler, 159
Taylor, A. E., 57, 83, 140
Tennant, 58, 120, 122, 206, 234, 268
Tertullian, 28, 175
Timaeus, 66, 69, 139 f., 191, 209
Traherne, T., 35
Tweedsmuir, Lord, 279 f.

Varro, 28, 33
Victorinus, 120

Index

Von Hügel, 156, 321

Ward, J., 180
Webb, C. C. J., 119, 292 f., 320
Wells, H. G., 245
Whichcote, 45
Whitehead, 29, 60, 104, 129, 160, 201, 260
Wisdom, J., 106 f.
Wittgenstein, 109, 189
Wollaston, 45

Worcester (Stillingfleet, Bishop of), 176

Xenocrates, 140
Xenophanes, 65

Yeats, 29, 125
York, Archbishop of, 37, 316

Zeno, 291
Zephaniah, 79